SELECTED READINGS IN PHYSICS

General Editor D. TER HAAR

Nuclear Forces

NUCLEAR FORCES

D. M. BRINK, D.Phil.

Fellow of Balliol College
University of Oxford

PERGAMON PRESS

OXFORD · LONDON · EDINBURGH · NEW YORK
PARIS · FRANKFURT

Pergamon Press Ltd., Headington Hill Hall, Oxford
4 & 5 Fitzroy Square, London W.1

Pergamon Press (Scotland) Ltd., 2 & 3 Teviot Place, Edinburgh 1

Pergamon Press Inc., 122 East 55th Street, New York 22, N.Y.

Pergamon Press GmbH, Kaiserstrasse 75, Frankfurt-am-Main

Federal Publications Ltd., Times House, River Valley Rd, Singapore

Samcax Book Services Ltd., Queensway, P.O. Box 2720, Nairobi, Kenya

First edition 1965

Library of Congress Catalog Card No. 65-18379

Set in 10 on 12pt Times
and Printed in Great Britain
by The Whitefriars Press Ltd., London and Tonbridge

Contents

Preface

IN RECENT years physics has grown into a very complex subject and even in one particular branch like Nuclear Physics so much is known that it is impossible for a student at undergraduate level to cover the whole ground. There is the danger that in trying to learn too much the student may not understand anything well and may feel frustrated in his attempt to assimilate too many facts. In these circumstances it may be interesting for the student, in the later stages of a course on Nuclear Physics, to follow up some restricted topic, to discover the central ideas, to find out when and in what context they arose and developed and to get an idea of the state of knowledge at the present time.

Our volume on "Nuclear Forces" has been compiled with this aim in mind. It might also be of interest to a more general reader who knows something of the basic facts of Nuclear Physics and who would like to learn something about the history and development of Nuclear Forces.

The book contains a collection of original papers that mark important stages in the development of ideas on Nuclear Forces and an introduction in which the ideas underlying the papers and subsequent events and theories are discussed. Four papers originally written in German have been translated into English. The size of the book made it necessary to select papers carefully, and it was unavoidable that some important works were omitted in the reprint section. The criteria adopted for the selection of the papers were that they should be important and interesting, that they should not be too difficult neither too long. These last two points have excluded several relevant papers such as the ones by White (1936) and Tuve, Heydenberg and Hafstad (1936) which contain the first evidence of a strong attractive nuclear force between protons and Wigner's (1937) paper which derives some of the consequences of the charge independence hypothesis.

Most of our reprints are of early papers. This is partly because the fundamental principles were found early in the history of the subject, partly because more recent theoretical papers are too difficult to be read by an undergraduate. Developments made during the last decade have been summarized in Part 1 which, it is hoped, will also provide a guide for further reading of original papers.

Two books on the Nucleon–Nucleon Interaction have been published lately by M. Moravcsik (1963) and R. Wilson (1963) respectively. These books aim to give a fairly complete picture of the current status of the subject and are both more advanced and more specialized than the present volume. Moravcsik treats the problem of nucleon–nucleon scattering theoretically while Wilson gives an account of experimental and phenomenological aspects. A reader wishing to make himself familiar with recent advances in Nuclear Forces as related to nucleon–nucleon scattering will find these two books very useful.

The author would like to thank the publishers of the following journals for permission to reprint original articles: *The Physical Review*, the *Zeitschrift für Physik*, *The Proceedings of the Physico-Mathematical Society of Japan*, *The Proceedings of the Royal Society*, *London*, *The Proceedings of the Chemical Society*, *London* and *Nature*, and also the authors of the various articles for confirming this permission. I am grateful to Professor W. Heisenberg for a discussion on the early history of the subject and to Professor T. Lauritsen for discussions on charge independence as well as for reading part of the manuscript. Finally I would like to thank my wife who translated four papers originally written in German, typed the manuscript, drew the figures and helped in many other ways with the preparation of the book.

D. M. B.

Part 1

I

Nuclear Physics in 1932

In 1931 Niels Bohr was invited to give the Faraday Lecture to the Chemical Society in London. In this lecture he reviewed the situation in atomic and nuclear physics at that time and exposed the many puzzling problems that presented themselves in the theory of nuclear structure. No answer could be given to them then. A great deal of information had been gathered about nuclei through a study of radioactivity, α-particle scattering, mass defect measurements and nuclear reactions. It was recognized that forces much stronger than electromagnetic ones acted between nuclei but nothing was known in detail of their nature.

The year 1932 was a turning point in the development of nuclear physics: The neutron was discovered and accelerators were used for the first time in the study of nuclear processes. On April 28th, 1932, the Royal Society (cf. Part 2, p. 121) held an important meeting on the structure of atomic nuclei. Lord Rutherford delivered the opening address reviewing the progress made in nuclear physics since the previous meeting on the same subject in February 1929. In this talk he spoke of the discovery of the neutron by Chadwick a few months earlier and announced the results of the first nuclear disintegration experiments with artificially accelerated ions carried out by Cockcroft and Walton (1932). Chadwick then described the discovery of the neutron. In the second part of this book we reproduce the addresses of Rutherford and Chadwick and also Bohr's lecture to the Chemical Society.

1.1 Nuclear Masses

It was known that many elements consisted of a mixture of isotopes and that the atomic weight of every isotope was very nearly an integer. The isotopic constitution of an element and the atomic weights of single isotopes could be determined with the mass spectrograph. In 1932 mainly through the work of Aston (1927) the atomic weights of about 250 isotopes were known. The mass spectrograph showed that the atomic weight of every nucleus was approximately an integer called the mass number A. The difference between the exact atomic weight M and the mass number was called the mass defect. The mass defects were interpreted by a model which assumed that the nucleus was composed of protons and electrons. The idea as stated by Rutherford at a Royal Society meeting in 1929 was as follows: "The difference in mass between the free and nuclear proton is ascribed to the packing effect, i.e. to the interaction of the electromagnetic fields of the protons and electrons in the highly condensed nucleus. On modern views we know that there is a close relation between mass and energy (Einstein's relation $E = mc^2$). The free proton has a mass 1·0073 while the proton in the nucleus has a mass very nearly 1. This apparently small loss of mass means that a large amount of energy has been emitted in the entrance of the free proton into the structure of the nucleus, an amount corresponding to about 7 MeV" (cf. Rutherford (1920), p. 395, Sommerfeld (1919)).

The binding energy per proton is almost constant throughout the entire periodic table which follows from the fact that Aston's packing fraction is very nearly constant except for the lightest elements. The significance of this important fact was probably first recognized by Gamow (1930). We shall discuss his contribution in Section 1.2.

1.2 Nuclear Radii

In 1911 Rutherford showed in his fundamental paper on atomic structure that nuclei had a radial extension of less than 10^{-12} cm. Rutherford's calculations were based on experiments of Geiger and Marsden in which α-particles were scattered by gold atoms in a thin foil. He showed that the force between the gold nucleus and the α-particle was the same as the force between two point charges, i.e. the Coulomb force, provided the separation of the two particles was greater than 10^{-12} cm. Scattering that follows this law is called "Coulomb scattering" or "Rutherford scattering".

Rutherford observed, in 1919, for the first time a possible departure from the "Rutherford scattering" law in an experiment in which α-particles were scattered from hydrogen atoms. This result was confirmed by the careful investigations of Chadwick and Bieler in 1921. Their results showed that the Coulomb force law between the α-particle and the proton held for separations greater than about 4×10^{-13} cm but became very much stronger for smaller separations. This experiment pointed to the existence of strong "nuclear forces" acting between nuclei. For many years, however, attempts were made to interpret them as some sort of electrical polarization forces.

A few years later in 1924 deviations from Rutherford's law were observed by Bieler in the scattering of α-particles by magnesium and aluminium. These results could be explained if the Coulomb law broke down when the separation of the centres of the nucleus and α-particle was about $3{\cdot}4 \times 10^{-13}$ cm. This distance was taken to be the radius of the nucleus. Similar experiments were performed during the next few years but deviations from Rutherford's law could be found only in light nuclei with the α-particle sources available at that time.

In 1927 Rutherford drew attention to the following paradox: Experiments on the scattering of α-particles by heavy nuclei showed no departure from Rutherford's scattering law. Thus the radii of these nuclei seemed to be smaller than $3{\cdot}2 \times 10^{-12}$ cm. On the other hand the speed of the slowest α-particle spontaneously emitted from the uranium nucleus required its radius to be at least $6{\cdot}5 \times 10^{-12}$ cm on classical ideas. Unless Coulomb's inverse square law of repulsion broke down for separations of the α-particle and the uranium nucleus equal to this value the α-particle would be retained inside the nucleus by a potential barrier and α-decay would not occur. Gamow (1928) and independently Condon and Gurney (1928) explained this paradox by pointing out that in wave mechanics a particle may penetrate a potential barrier. Quantitative calculations gave values for the radii of heavy radioactive nuclei which were all of the order of 8×10^{-13} cm.

We know nowadays that the density distribution of matter in a nucleus is approximately constant in its interior and falls rapidly to zero at the nuclear surface. Recent experiments show that the radii of all nuclei may be represented approximately by a formula

$$R = 1{\cdot}2A^{\frac{1}{3}} \times 10^{-13} \text{ cm}$$

where A is the mass number. The volume of the nucleus is approximately proportional to A or equivalently, the density of "nuclear matter" in the interior of nuclei is nearly constant throughout the whole periodic table.

The position was not nearly so clear in 1932. A certain amount was known about the size of light nuclei from the scattering of α-particles and about heavy nuclei from α-decay life times. As early as November 1928 Gamow had suggested that the nuclear radius was proportional to $A^{\frac{1}{3}}$ or that the nuclear volume was proportional to A. He pointed out that Bieler's (1924) value for the radius of the aluminium nucleus ($R = 3{\cdot}4 \times 10^{-13}$ cm) and his own values for the radii of heavy nuclei (e.g. RaEm, $R = 7{\cdot}35 \times 10^{-13}$ cm) were consistent with the hypothesis that the nuclear radius R is related to the mass number A by the approximate relation

$$R = 1{\cdot}2A^{\frac{1}{3}} \times 10^{-13} \text{ cm}$$

There were hardly sufficient data to prove Gamow's hypothesis—just two points on the curve—but the idea was consistent with the α-particle model of the nucleus he put forward two months later at the Royal Society discussion in February 1929 (Gamow 1929). In this theory he assumed that the nucleus could be treated as a collection of α-particles which behaved like hard spheres interacting by strong attractive forces decreasing rapidly with distance as well as by relatively weak Coulomb forces. Such a collection of α-particles would have properties similar to a small drop of liquid in which the particles are held together by surface tension. On the basis of this model one would expect that the volume of the nucleus should be proportional to the number of α-particles contained in it. One would also expect the total binding energy (neglecting Coulomb forces and surface energy) to be proportional to the number of α-particles, a result which is consistent with mass defect measurements. Gamow's hypothesis that the nuclear volume was proportional to the mass number was generally accepted in 1932 and Heisenberg used it in his discussion of nuclear stability (cf. p. 150).

Majorana in his paper in 1933 on the proton–neutron model of the nucleus says: "Thus one finds at the centre of an atom a sort of matter which has the same property of uniform density as ordinary matter. Light and heavy nuclei are built up from this matter and the difference between them depends mainly on their different content of nuclear matter." Even though Gamow's α-particle model of the nucleus was soon to be superseded by the proton–neutron model it was important because it introduced the idea of nuclear matter. This was necessary for the subsequent theories of Heisenberg and Majorana. The constant density of nuclear matter and the proportionality of the binding energy to the mass number A are important facts which impose some conditions on nuclear forces. Forces that reproduce these properties satisfy the *saturation condition.*

1.3 Spin and Statistics

At the Royal Society discussion on nuclear structure in April 1932 Fowler stressed the puzzle of the spin and statistics of N^{14}. It was commonly believed that nuclei were made up of protons and electrons with α-particles as subunits in the structure. One weakness of this theory was the incorrect prediction of the spin and statistics of N^{14}.

It was known that both protons and electrons obeyed Fermi statistics and had spin angular momentum of $\frac{1}{2}\hbar$. There were two rules of thumb concerning the spin and statistics of a composite system made up of particles with spin $\frac{1}{2}\hbar$ obeying Fermi statistics:

(i) The composite system obeys Bose and Fermi statistics depending on whether the number of component particles is even or odd. This rule was proved in 1931 by Ehrenfest and Oppenheimer.

(ii) The composite system has an angular momentum which is an integral or half-integral multiple of $\hbar$ depending on whether the number of component particles is even or odd.

If N^{14} is made up of 14 protons and 7 electrons these rules predict that it should obey Fermi statistics and have an angular momentum which is a half-odd integral multiple of $\hbar$. In fact N^{14} has a spin of $\hbar$ and obeys Bose statistics. The spin of N^{14} had been measured by Ornstein and Van Wijk in 1928. Kronig (1928) pointed out that the experimental result was inconsistent with rule (ii) if the nucleus was composed of protons and electrons. Heitler and Hertzberg (1929) showed that the statistics were also inconsistent with rule (i) after the Raman spectrum of the N^{14} molecule had been studied by Rassetti (1929).

The lowest energy levels of the homonuclear N_2 molecule form a rotational band with states labelled by the orbital angular momentum quantum number L. Each orbital state has a certain spin degeneracy or statistical weight which depends on the spin and statistics of the nitrogen nucleus. The total nuclear spin of the nitrogen molecule is the sum of the spins of the two individual nuclei. If the nuclei spin is I the number of even spin states of the two nuclei in the molecule is $(I+1)(2I+1)$ and the number of odd spin states is $I(2I+1)$. If the nitrogen nucleus obeys Fermi statistics the total wave function must be antisymmetric for exchange of the two nuclei by the Pauli principle. Thus if the orbital angular momentum L is even the spin symmetry must be odd and vice versa. States with L even would have a statistical weight of $I(2I+1)$ and the states with L odd a statistical weight of $(I+1)(2I+1)$. If the nitrogen nucleus obeys Bose statistics the statistical weights are interchanged. Raman transitions obey a selection rule $\Delta L = \pm 2$ and an investigation of the mechanism of the process shows that the intensity of lines in the Raman spectrum is proportional to the statistical weights of the states involved. Thus lines in the spectrum alternate in intensity with an intensity ratio $(I+1)/I$. If the lines corresponding to states with even L are stronger the nuclei obey Bose statistics. If the odd lines are stronger they obey Fermi statistics. The intensities of the Raman lines give both the spin and statistics of the nitrogen nucleus.

Rassetti made the first measurements of the Raman spectrum of N_2 in 1929. He remarked that although H_2 and N_2 have a similar electronic structure they behave in opposite ways with regard to the relative weights of odd and even rotational states. Two months later Heitler and Hertzberg commenting on Rassetti's results pointed out that his measurements implied nitrogen obeyed Bose statistics, contrary to expectations based on the proton–electron model of the nucleus.

1.4 The Continuous β-ray Spectrum

The energy spectrum of β-particles emitted from radioactive nuclei is complex. Most often there are groups of β-particles with definite energy superimposed on a background with a continuous energy distribution. RaE is an exception to this general rule in that it has only a continuous spectrum. The groups with definite energy were first noticed in 1911 by Bayer, Hahn and Meitner and later, in 1914, Chadwick discovered a continuous background. Rutherford and his co-workers showed that the discrete energy groups were of secondary origin produced by conversion of γ-rays. They proposed that γ-rays originating in the nucleus were absorbed by atomic electrons and ejected these from their orbits with a definite energy equal to the γ-ray energy minus the binding energy of the electron in its orbit round the atom. The primary β-rays had the continuous energy spectrum discovered by Chadwick.

In 1922 Meitner pointed out that a nucleus, presumably quantized, ought not to emit electrons of varying energy and suggested that the inhomogeneity was introduced after the actual emission of the β-ray from the nucleus. Experiments produced no evidence in favour of this idea and one performed by Ellis and Wooster (1927) even contradicted it. They found the average amount of energy per disintegration by measuring the heat produced when a known number of atoms decayed in a calorimeter so thick that no β-rays could escape from it. They chose RaE for their experiment because it had only a continuous spectrum and it was known that there was no penetrating γ-radiation produced in conjunction with the β-rays.

If Meitner's suggestion was true and all decays produced the same energy, then atoms emitting slower electrons must get rid of their energy in some other way. Ellis and Wooster knew that there was no large amount of penetrating γ-radiation produced and thus the surplus energy should be absorbed in the calorimeter. The decay energy must be at least as large as the maximum β-ray energy (1·05 MeV for RaE). So the average energy per decay measured by the calorimeter method should be at least 1·05 MeV. If, on the contrary, the disintegration electrons were ejected from the nucleus with various energies, then the average energy per decay should correspond to the mean energy of the continuous spectrum (i.e. $0{\cdot}39 \pm 0{\cdot}06$ MeV). Ellis and Wooster found the average energy per decay to be $0{\cdot}35 \pm 0{\cdot}04$ MeV, supporting the second hypothesis and providing strong evidence against the first.

This property of β-decay was in striking contrast to the known features of α-disintegration. Early experiments had shown that the α-particles from a given nucleus were all emitted with about the same energy (except for the long-range α-emitters). In 1930 the fine structure of the α-spectrum was discovered, and it was found that α-particles were emitted in groups, each group with a definite energy. Also, the energy differences between the α-groups seemed to correspond very closely to the energies of γ-rays produced in association with the α-decay (cf. Part 2, p. 125). These results suggested that the α-decay led sometimes to excited quantum states of the final nucleus which decayed subsequently emitting γ-rays and they supported the application of the conservation of energy to nuclear decays producing α-rays and γ-rays.

Niels Bohr discussed the current problems of nuclear physics in his Faraday Lecture in 1931 (cf. Part 2, p. 138) and again at the Rome conference of nuclear physics in October 1931. On both occasions he was concerned about the theoretical implications of the peculiar features of β-decay and especially about the principle of energy conservation. After a careful discussion of the evidence he concluded (cf. Part 2, p. 143): "At the present stage of atomic theory, however, we may say that we have no argument, either empirical or theoretical, for upholding the energy principle in the case of β-ray disintegrations and are even led to complications and difficulties in trying to do so."

The principle of energy conservation was saved by the neutrino theory of β-decay. Pauli suggested a hypothesis that the law of conservation of energy remained valid and the expulsion of β-particles was accompanied by a very penetrating radiation of neutral particles which had not been observed. The sum of the energies of the β-particle and the neutral particle (or particles if there were more than one) emitted by the nucleus in each decay would be equal to the upper limit of the β-ray energy spectrum, and the neutral particles would have escaped from the calorimeter in the experiment by Ellis and Wooster and would not have contributed to the heating effect.

Pauli first mentioned this hypothesis in 1930 in a letter to Geiger and Meitner at Tübingen University. He made his suggestion public at a meeting of the American Physical Society in Pasadena in June 1931 and probably mentioned it in discussions at the Rome conference in October 1931. The first published statement of the hypothesis appeared in the discussion of Heisenberg's paper on the proton–neutron model of the nucleus at the Solvay conference in October 1933. Fermi was present at that conference and published his theory of β-decay based on Pauli's hypothesis a few months later. There are fuller accounts of the early history of the neutrino in a lecture by Pauli (1957), by C. S. Wu (1960) in a Memorial Volume to Pauli and by F. Rassetti (1962) in the Collected Works of Fermi. The name "neutrino" was suggested by Fermi.

We give these historical details in order to have a picture of the background against which Heisenberg's theory of nuclear structure was developed. Heisenberg published his first paper on the subject in June 1932. By that time Pauli's hypothesis was known to many physicists, but it was not yet universally accepted. Bohr was certainly not convinced of its validity when he gave his lecture in Rome in October 1931 (both Pauli and Heisenberg being present). Judging by Heisenberg's first paper it would seem that he still had an open mind as to a possible breakdown in the energy principle when he wrote it. The published reports of discussions at the Solvay conference show, however, that Heisenberg and others gave considerable support to Pauli's idea by the end of 1933.

One of the remarkable features of Heisenberg's first studies of nuclear structure was that he made them at a time when the foundation of all the established theory seemed in doubt. He succeeded in his programme by avoiding a discussion of the mechanism of β-decay and applying the established theory to the motion of protons and neutrons in the nucleus.

1.5 The Neutron and the Positron

Before the neutron was discovered there were many speculations predicting its existence. This hypothetical particle was generally conceived to be a tightly bound combination of a proton and an electron and there were suggestions that it was a component in the structure of heavier nuclei (Rutherford 1920). In February 1932 Chadwick published a letter in *Nature* giving evidence for the existence of the neutron and at the discussion on Nuclear Structure at the Royal Society in April (cf. p. 133) he reviewed the events leading up to his experiments and discussed the results. The neutron was first observed in radiation produced by the action of α-rays on beryllium and boron. Chadwick showed that the experimental results could be explained assuming the radiation consisted of particles with zero charge and a mass between 1·005 and 1·008 atomic mass units. This value for the neutron mass was rather smaller than the one accepted today (1·00898 a.m.u.). It was mainly due to inaccuracies in Aston's mass spectrograph data (1927). These showed up later as inconsistencies between mass differences determined by the mass spectrograph and nuclear reaction data.

Heisenberg, who was working in Leipzig at the time, submitted his fundamental paper on the neutron–proton model of the nucleus in June 1932. In the introduction he pointed out that the problem of the spin and statistics of N^{14} could be resolved if the neutron obeyed Fermi statistics and had a spin $= \frac{1}{2}\hbar$. This argument is sufficient to show that the neutron spin is a half-odd integral multiple of $\hbar$. Heisenberg chose $\frac{1}{2}\hbar$ for reasons of simplicity. The deuteron spin which is equal to $\hbar$ restricts the neutron spin to $\frac{1}{2}\hbar$ or $\frac{3}{2}\hbar$, but the first direct evidence excluding the possibility of $\frac{3}{2}\hbar$ was not published until 1937. It comes from experiments on neutron scattering from ortho- and para-hydrogen (Schwinger, 1937).

In 1930 it was found that hard γ-rays were absorbed much more rapidly in heavier elements than was expected and secondary γ-rays were observed to be given off in the absorption process. These could be resolved into two components with energies of about 0·5 MeV and 1 MeV respectively (cf. Part 2, p. 127, p. 154). The origin of secondary γ-rays was not explained for several years and was still a source of misunderstanding when Heisenberg developed his first ideas about nuclear structure in 1932. The difficulties were resolved after Anderson's discovery of the positron (Sept. 1932), and it was soon recognized that hard γ-rays interacting with matter produced electron-positron pairs, and that a positron could annihilate with an electron in matter producing two 0·5 MeV γ-rays or one 1 MeV γ-ray.

1.6 Accelerators

In his address to the Royal Society meeting in 1932 Rutherford described the development of positive ion accelerators by groups of research workers at the Cavendish Laboratory in Cambridge, the Department of Terrestrial Magnetism in Washington and the University of California in Berkeley and announced the results of the first experiments using accelerators for nuclear studies carried out at Cambridge. The other two groups performed similar experiments within several months.

The development of accelerators was a major event in nuclear physics and was one of the main reasons for the rapid advances made during the next few years. Up to this time most information on the structure of the nucleus had come from experiments with α-particles, but accelerators formed an additional line of attack which promised to have many advantages. The intensity of the source would be much greater, as 1μA of positive ions was equivalent to 180 g of radium, the beam of particles would be free from β and γ rays, which caused complications in many experiments and the energy of the particles could be varied at will.

Cockcroft and Walton began their work at the Cavendish Laboratory in 1929. In 1930 using a half-wave rectifier to obtain a constant potential and a single section accelerator tube they produced protons with an energy of 300 keV. Next they developed a more powerful machine with a voltage multiplier and a two-section tube and obtained protons with energies up to 700 keV. The first disintegration experiments performed with this machine on the reaction

$$Li^7 + p \rightarrow 2\,He^4$$

were announced at the Royal Society discussion on Nuclear Structure in April 1932.

E. O. Lawrence of the University of California began to work on the design of the cyclotron in 1929. In 1930 he had a small model working and obtained protons with an energy of 80 keV. By 1931 Lawrence and Livingston had constructed a larger machine and produced protons with energies up to 1 MeV. The results of the first experiments on nuclear studies were published in 1932. These experiments extended the work of Cockcroft and Walton up to 910 keV.

Earlier, in 1926, G. Breit had started investigations of the Tesla coil as a means of producing high voltages. He was joined by Tuve, at the Department of Terrestrial Magnetism in Washington, and in the years up to 1930 they made great progress in the design of accelerator tubes. But the Tesla coil was not satisfactory as a high voltage source because of the pulsed nature of the potential and the fluctuating value of the peak voltage. At Princeton in 1929, R. G. Van de Graaff built the first electrostatic generator of the type named after him. At a very early stage Tuve and his co-workers realized the virtues of the Van de Graaff generator as a high voltage source, and in 1932 they built a small model provided with a sectional metal and glass accelerating tube which had been used previously in the Tesla coil work. In 1933 they published results of nuclear disintegration experiments using ions accelerated by this machine with energies of up to 600 keV. This was the first successful application of the Van de Graaff generator to nuclear studies.

II

The Theories of Heisenberg Wigner and Majorana

THE discovery of the neutron in 1932 led to renewed thinking about nuclear structure. The possible existence of neutrons had been conjectured earlier by Rutherford (1920), when he had discussed the possibility of heavy nuclei being formed of protons and neutrons. It seemed natural now to investigate this idea more seriously after the existence of neutrons had been demonstrated by experiment. Soon several physicists, Iwanenko (1932), Chadwick (1932), Heisenberg (1932) pointed out that the difficulty of the spin and statistics of N^{14} (cf. Chap. I, Section 5) would be removed, if nuclei were composed of protons and neutrons, and if the neutron had a half-integral spin and obeyed Fermi statistics.

Three suggestions on the nature of the forces acting between nuclear particles became most important for the subsequent development of nuclear studies. Shortly after the neutron was discovered, Heisenberg, Wigner and Majorana published papers on this problem. Each of these three authors made specific postulates regarding the nature of nuclear forces and studied the consequences for some features of nuclear structure. The picture of the nucleus they proposed has remained essentially unchanged since then, and the three types of forces, Heisenberg-, Wigner- and Majorana-forces, have been widely used in calculations. We discuss the three papers in this chapter and the original texts are reproduced in Part 2.

2.1 Heisenberg's Exchange Interaction

In a series of three papers Heisenberg put forward a number of ideas which formed the basis for later investigations of nuclear forces and nuclear structure. In the reprint section of this book we reproduce the first of these papers and part of the third which is relevant to a discussion of nuclear forces.

In his papers Heisenberg suggested that nuclei were composed of protons and neutrons and that nuclear structure could be described by the laws of quantum mechanics in terms of the interaction between the nuclear particles. He submitted the first paper in the series in June 1932, just four months after Chadwick had published his discovery of the neutron.

Nowadays we believe that both neutrons and protons are elementary particles. Heisenberg expressed this view at the beginning of his first paper and used it as a basis for his discussion of nuclear structure. On the other hand, he wished to suggest a possible mechanism for the interaction between nuclear particles, and for this purpose he still retained the idea that neutrons were made up of protons and electrons in some sort of close combination. This picture suggested an exchange interaction between protons and neutrons analogous to the resonance interaction between a hydrogen H^+-ion and a hydrogen atom. Because this analogy was so important for the subsequent development of ideas about exchange interactions we give a brief review of the problem of a hydrogen atom interacting with a H^+-hydrogen ion by the exchange of an electron.

We have to investigate the motion of one electron in the field of two hydrogen nuclei fixed at points A and B in space. We assume that the separation between A and B is rather larger than the radius of the hydrogen atom so that the interaction is weak and simple approximation methods have some validity. There are two approximate solutions to the problem of an electron moving around the two nuclei both having the same mean energy. The first with a wave function ϕ_A corresponds to a hydrogen atom at A undisturbed by the hydrogen nucleus at B and the second with wave function ϕ_B describes a hydrogen atom at B and a hydrogen nucleus at A. The off-diagonal matrix elements of the Coulomb interaction of the electron with the two nuclei splits the degeneracy between these two states, and the eigenfunctions ϕ_1 and ϕ_2 may be represented approximately by linear combinations of the wave functions ϕ_A and ϕ_B. These combinations must be either symmetric or antisymmetric because of the symmetry of the force field around the two nuclei.

$$\phi_1 = B_1(\phi_A+\phi_B), \qquad \phi_2 = B_2(\phi_A-\phi_B)$$

Here B_1 and B_2 are normalization constants. If the separation between A and B is large so that the overlap of ϕ_A and ϕ_B is small, $B_1 \simeq B_2 \simeq 1/\sqrt{2}$. The states ϕ_1 and ϕ_2 have energies $-J(r)$ and $J(r)$ respectively ($J(r) > 0$) where $J(r)$ depends on the separation r of the two nuclei A and B. Thus the residual interaction between the atom and ion is attractive or repulsive depending on the symmetry of the electron wave function.

In Chap. VI we shall need a formula for the exchange energy $J(r)$. In order to calculate it let V_A and V_B be the Coulomb potentials of the nuclei A and B. If the kinetic energy of the electron is T, its total energy is $H = T + V_A + V_B$. The wave functions ϕ_A and ϕ_B satisfy the Schrödinger equations

$$(T+V_A)\phi_A = -\varepsilon\phi_A \qquad (T+V_B)\phi_B = -\varepsilon\phi_B \tag{2.1}$$

To calculate the exchange energy we find the mean energies $<E_1>$ and $<E_2>$ of the states ϕ_1 and ϕ_2

$$\begin{aligned}
\langle E_1\rangle &= \int \phi_1(r)H\phi_1(r)\,\mathbf{dr} \\
&= B_1^2 \int (\phi_A+\phi_B)(T+V_A+V_B)(\phi_A+\phi_B)\,\mathbf{dr} \\
&= B_1^2 \int (\phi_A+\phi_B)[(T+V_A)\phi_A+(T+V_B)\phi_B]\,\mathbf{dr}\, + \\
&\qquad + B_1^2 \int (\phi_A+\phi_B)(V_A\phi_B+V_B\phi_A)\mathbf{dr}
\end{aligned} \tag{2.2}$$

Using the Schrödinger equations (2.1) for ϕ_A and ϕ_B the first term of equation (2.2) reduces to $-\varepsilon$ and the second term is $K-J$ where

$$\begin{aligned}
J &= -B_1^2\left[\int \phi_A V_A \phi_B\,\mathbf{dr} + \int \phi_B V_B \phi_A\,\mathbf{dr}\right] \\
&= -2B_1^2 \int \phi_A V_A \phi_B\,\mathbf{dr}
\end{aligned}$$

because of the symmetry between A and B. Similarly

$$K = 2B_1^2 \int \phi_A V_B \phi_A\,\mathbf{dr}$$

If the separation between A and B is large $|K| \ll |J|$ and $B_1^2 \simeq \frac{1}{2}$. In this limit

$$\langle E_1 \rangle = -\varepsilon - J, \quad \text{and} \quad \langle E_2 \rangle = -\varepsilon + J$$

and the exchange integral

$$J \simeq -\int \phi_A V_A \phi_B \,\mathrm{d}\mathbf{r} \tag{2.3}$$

If a hydrogen atom is placed at a point A near a hydrogen ion at B the electron will move from the ion A to the ion B and back again. In order to do so, it has to penetrate a potential barrier because the total energy of the electron is less than its potential energy halfway between the ions, unless the ions are very close together. The strength of the interaction between the ions is related to the probability that the electron penetrates this barrier and jumps across from A to B. We recall that the wave function of a stationary state with energy E has a time dependence $\exp(-iEt/\hbar)$ (where $\hbar = \mathbf{h}/2\pi$). We now consider a state $\phi(t)$ which is a superposition of the stationary states ϕ_1 and ϕ_2 defined by the equation

$$\begin{aligned} \phi(t) &= B_1[\phi_1 \exp(iJt/\hbar) + \phi_2 \exp(-iJt/\hbar)] \\ &= \phi_A \cos(J(r)t/\hbar) + i\phi_B \sin(J(r)t/\hbar) \end{aligned}$$

When $t = 0$ this wave function is equal to ϕ_A and represents the initial situation of a hydrogen atom at A and an ion at B. When $Jt/\hbar = \pi/2$, $\phi(t) = \phi_B$ and represents an ion at A and an atom at B. Thus the electron oscillates between the ions through the potential barrier that separates them with a frequency $J(r)/\mathbf{h}$. The exchange frequency is proportional to the interaction energy.

In addition to the exchange interaction between neutrons and protons Heisenberg included a weaker interaction between neutrons. He also assumed the Coulomb repulsion to be the only force acting between two protons. This seems to have been a consequence of his picturing the proton as an elementary and the neutron as a composite particle. In later work Majorana and Wigner dropped the neutron–neutron interaction and argued that the Coulomb repulsion could be neglected in light nuclei. For a number of years most investigations assumed the nuclear interaction between like particles to be non-existent or very weak compared with the neutron–proton interaction. This assumption was proved wrong by the experimental investigations on proton–proton scattering by White and by Tuve, Heydenberg and Hafstad in 1936. Subsequently, the strength of the nuclear interaction between like nuclear particles was found to be almost equal to the neutron–proton interaction strength. This property of nuclear forces is called *charge-independence* and we discuss it more fully in Chap. IV.

In his first two papers Heisenberg used his ideas to find qualitative rules governing the stability of nuclei against α- or β-decay. In his third paper published in December 1932 he attempted to find an approximate solution of the equations of motion of protons and neutrons in the nucleus. This work revealed a difficulty: Heisenberg applied the Thomas-Fermi approximation method to nuclei and showed that the nuclear binding energy as a function of atomic weight A increased more rapidly than A^2 if the strength of the exchange interaction $J(r)$ was positive for all values of r. Empirically, Aston's mass defect measurements showed that the binding energy was approximately proportional to A if A was small, and for large A it increased even more slowly. Thus, Heisenberg's exchange interaction did not produce saturation. He found that the same conclusion held for an ordinary attractive potential interaction. In order to obtain saturation he returned to Gamow's liquid drop model and assumed that the force between the nuclear particles became strongly repulsive if they approached each other more closely than a certain critical distance. In other words, he postulated that the interaction between nuclear particles had a repulsive core. Another way out of the difficulty was proposed by Majorana in March 1933 and will be discussed in Section 4 of this chapter.

2.2 The Isobaric Spin Formalism

It is interesting that already in his first paper on the proton–neutron theory of nuclear structure Heisenberg introduced the ρ-spin or, as it is called now, the isobaric (isotopic) spin formalism. He treated the proton and neutron as two states of the same nuclear particle or *nucleon* which were distinguished by the quantum number ρ^ζ. This quantum number had the value $+1$ when the nucleon was a neutron and -1 when it was a proton. Nowadays another notation is usual with Heisenberg's ρ replaced by τ. Also the most common sign convention is opposite: $\tau^\zeta = +1$ means the nucleon is a proton and $\tau^\zeta = -1$ that it is a neutron (cf. p. 146).

The isobaric spin formalism was not strictly necessary for Heisenberg's theory of nuclear structure, and in a lecture at the Solvay conference in October 1933 he showed that it was equivalent to a formalism that treats neutrons and protons as different particles. (We will call this the ordinary method.) The isobaric spin method was used later by Fermi (1934) in his theory of β-decay and by Yukawa (1935) in his meson theory of nuclear forces, but in those early years physicists felt it was too complicated and preferred the ordinary method. For example Majorana in his paper in 1933 was pleased that he could avoid those "troublesome ρ-spin coordinates". After the charge independence of nuclear forces had been postulated in 1936, the isobaric spin formalism was revived and became the natural mathematical representation of this physical symmetry. We will discuss the two methods and compare the wave functions for a system consisting of a proton and a neutron:

(i) In the ordinary method each proton (neutron) has four coordinates, three position coordinates $\mathbf{r}_p$ ($\mathbf{r}_n$) and one spin coordinate σ_p (σ_n). To simplify the notation we shall sometimes denote the four coordinates ($\mathbf{r}_p$, σ_p) of the proton by one letter u_p (u_n for the neutron). The wave function of the neutron–proton system is some function $\psi(u_p, u_n)$ of these coordinates.

(ii) In the isobaric spin method each particle has five coordinates, the spin and space coordinates $(\mathbf{r}_k, \sigma_k) = u_k$ and a new coordinate τ_k with the value $+1$ for a proton and -1 for a neutron. The two-particle system is characterized by the coordinates $(u_1, \tau_1; u_2, \tau_2)$. There is, however, an ambiguity in that the suffices 1 and 2 have no physical significance and the two sets of coordinates $(u_1, \tau_1; u_2, \tau_2)$ and $(u_2, \tau_2; u_1, \tau_1)$ both describe the same physical situation. This ambiguity is removed if the wave function has the same numerical value except for a constant phase factor for these two sets of coordinates. A consistent scheme is obtained if all wave functions are required to satisfy the extended Pauli principle, i.e. if they must be antisymmetric for the exchange of all five coordinates of any two particles. In a two-particle system we have therefore

$$\phi(u_1, \tau_1; u_2, \tau_2) = -\phi(u_2, \tau_2; u_1, \tau_1)$$

In our example the system consists of a proton and a neutron. Hence either particle 1 is a proton and particle 2 a neutron ($\tau_1 = 1$, $u_1 = u_p$ and $\tau_2 = -1, u_2 = u_n$) or vice versa ($\tau_1 = -1, u_1 = u_n$ and $\tau_2 = 1, u_2 = u_p$). The wave functions in the two schemes are related by the equation

$$\phi(u_1, +1; u_2, -1) = -\phi(u_2, -1; u_1, +1) = \psi(u_1, u_2)$$
$$\phi(u_1, \tau_1; u_2, \tau_2) = 0 \quad \text{if } \tau_1 = \tau_2 = \pm 1.$$

This equation establishes a one-to-one correspondence between wave functions in the two representations. In concluding this discussion, it should be emphasized that the isobaric spin method is completely equivalent to the ordinary method. It is just another way of describing a system made up of protons and neutrons and requires no physical assumptions other than that neutrons and protons both obey Fermi statistics.

Heisenberg's exchange interaction between a pair of nucleons k and l is contained in the second term of his Hamiltonian function and may be written as

$$J(r_{kl})(P_c')_{kl}$$

where $r_{kl} = |\mathbf{r}_k - \mathbf{r}_l|$ is the distance between the nucleons, $J(r_{kl})$ is the strength of the exchange interaction, and the operator

$$(P_c')_{kl} = \tfrac{1}{2}(\tau_k^\xi \tau_l^\xi + \tau_k^\eta \tau_l^\eta)$$

is a charge exchange operator.† If the nucleons k and l are both protons or both neutrons this operator is equal to zero. If one is a

† The exchange operators used in this section are slightly different from the conventional ones in that they exchange certain coordinates of a pair of unlike particles but give a zero result for a pair of like particles. We attach a prime to the exchange operators in this section to distinguish them from the conventional ones.

proton and the other a neutron it interchanges their isobaric spin coordinates τ_l and τ_k and so converts the proton into a neutron and vice versa. More precisely, if $\phi(\ldots, u_k\tau_k, u_l\tau_l, \ldots)$ is a many-particle wave function in the isobaric spin representation, the charge exchange operator may be defined by the equations

$$(P'_c)_{kl}\phi(\ldots, u_k\tau_k, u_l\tau_l, \ldots) = 0 \quad \text{for } \tau_k = \tau_l$$
$$= \phi(\ldots, u_k\tau_l, u_l\tau_k, \ldots) \quad \text{for } \tau_k \neq \tau_l$$

In his lecture at the Solvay conference Heisenberg showed that his charge exchange interaction was equivalent to a combined space and spin exchange interaction. We illustrate this by a diagram in Fig. 1,

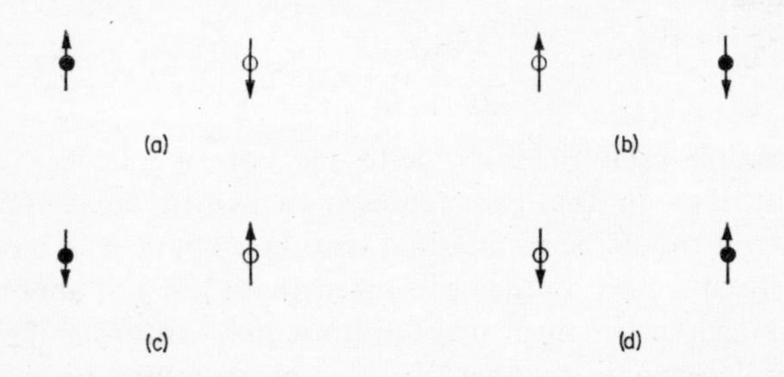

FIG. 1. Illustration of exchange processes. A dot represents a proton, a circle a neutron. The arrows indicate spin directions of the particles. (a) The original state: a proton and neutron with opposite spins, (b) shows the effect of charge exchange, (c) spin exchange, (d) exchange of position.

where a dot represents a proton and a circle a neutron. The arrows indicate the spin direction of the particles. Thus, Fig. 1a represents a proton and a neutron with opposite spins.

Figure 1b shows the effect of a charge exchange between the two nucleons, where the first particle changes to a neutron and the second to a proton while the spins remain unchanged. Similarly, Fig. 1c illustrates a spin exchange and Fig. 1d an exchange of position. We notice that any one of the exchange processes may be produced by a combination of the other two. In particular, a spin exchange followed by a position exchange is equivalent to a charge exchange.

The equivalence of charge exchange to spin exchange followed by position exchange may be proved analytically from the extended Pauli principle. If P'_M and P'_σ are the position and spin exchange operators for a proton–neutron pair, then the operator $(P'_M)_{kl}(P'_\sigma)_{kl}$ interchanges the position and spin coordinates u_k and u_l of the two nucleons. Thus

$$\begin{aligned}(P'_M)_{kl}(P'_\sigma)_{kl}\phi(\ldots u_k\tau_k, u_l\tau_l, \ldots) &= \phi(\ldots, u_l\tau_k, u_k\tau_l, \ldots)\\ &= -\phi(\ldots, u_k\tau_l;\ u_l\tau_k, \ldots)\\ &= -(P'_c)_{kl}(\ldots, u_k\tau_k;\ u_l\tau_l, \ldots)\end{aligned}$$

The second step in this calculation follows from the extended Pauli principle which requires the wave function ϕ to change sign if all five coordinates of two particles are interchanged and the third step follows from the definition of P'_c. The equation holds if k and l are a proton–neutron pair, i.e. if $\tau_k \neq \tau_l$. If the nucleons k and l are identical, all the operators P'_M, P'_σ, and P'_c are zero and the equation is an identity. So, the charge exchange operator

$$(P'_c) = -P'_M P'_\sigma$$

In the above discussion we used the isobaric spin formalism, but the position and spin exchange operators may be defined in the ordinary formalism which treats protons and neutrons as different particles. Since Heisenberg's charge exchange operator P'_c is equivalent to the operator $-P'_M P'_\sigma$ his charge exchange interaction can be written in the ordinary formalism by using the position and spin exchange operators, and all his calculations could be carried through without the isobaric spin formalism.

2.3 Wigner's Problem

One of the striking features of the mass defects of light nuclei is the very large binding energy of the α-particle (He^4). The binding energies of the deuteron and α-particle are 2·2 MeV and 28 MeV respectively. Thus, the binding energy of the α-particle is about thirteen times as big as that of the deuteron.

Wigner (1932) set out to find a simple explanation of this experimental fact. Assuming a potential $V(r)$ between a proton and a neutron depending on their separation r and neglecting forces between two protons and two neutrons, he showed that the binding energies of light nuclei could be estimated using standard quantum mechanical methods. He made calculations for the α-particle and the deuteron and showed that the binding energy of the α-particle would be much larger than that of the deuteron if the potential $V(r)$ had a sufficiently short range. In his conclusions Wigner pointed out the physical reason for the effect (cf. Part 2, p. 181): "The difference of the mass defects of He and H^2 can be attributed to the great sensitivity of the total energy to a virtual increase of the potential as is brought about by the fact that every particle in He is under the influence of two attracting particles instead of one as in the case of H^2." The reason for this sensitivity is that the deuteron is a very weakly bound structure with a mean potential energy almost equal and opposite to its mean kinetic energy and much greater than its binding energy. In such a situation a small relative change in the strength of the interaction produces a large relative change in the binding energy.

Wigner's theory of the mass defect of helium and Heisenberg's third paper on nuclear structure were both submitted for publication in December 1932 and neither Wigner nor Heisenberg knew of the other results at this stage. In the course of his discussion on nuclear binding energies Heisenberg proved that a force of the type considered by Wigner would not satisfy the saturation criterion hence it could not be responsible for nuclear binding. However, the physical idea underlying this explanation of the stability of the α-particle was very general and the arguments could be extended to prove the same result for some other interactions, particularly for the exchange interaction introduced by Majorana in March 1933.

It is amusing to find that five years after Wigner's argument was published theoretical physicists were trying to understand why the binding energy of He^4 was only 28 MeV and not larger. Wigner had neglected the forces between two protons or two neutrons, but in 1936 they were found to be as strong as the neutron–proton forces. Thus, there were six "bonds" in He^4 instead of only four, the potential energy was increased by a factor 3/2 and the binding energy by an even larger one. The tensor forces discovered in 1939 helped to reduce the theoretical He^4 binding energy to the experimental value.

2.4 Majorana's Exchange Interaction

Heisenberg was guided by an analogy when he was trying to find a suitable interaction between nuclear particles. He treated the neutron as a composite particle and assumed an exchange interaction between a proton and a neutron similar to the one responsible for the molecular binding of the H and H^+ ions.

Majorana doubted the validity of this analogy and pointed out that Heisenberg's theory did not explain the saturation of nuclear binding energies without the ad hoc assumption that nuclear forces became strongly repulsive for small separations. He preferred an alternative approach: He wanted to find the simplest law of interaction between nuclear particles which would produce saturation. At the beginning of his paper he rejected the idea that forces became strongly repulsive at short separations, on the grounds that "such a solution to the problem was aesthetically unsatisfactory, since one would have to assume not only attractive forces of unknown origin between elementary particles, but also, for small separations, repulsive forces of enormous magnitude corresponding to a potential of several hundred million MeV" (cf. Part 2, p. 163).

In Heisenberg's theory the exchange interaction strength $J(r)$ was taken to be positive for all values of r, the proton–neutron separation. This was a natural assumption, if one followed the analogy between nuclear and molecular forces. The interaction, however, did not produce saturation. Majorana made his first important contribution when he noticed that Heisenberg's exchange interaction would satisfy the saturation criterion without a repulsive core if the sign of the interaction strength was changed, i.e. if $J(r) < 0$ for all r.

In Section 3 of this chapter we showed that Heisenberg's exchange interaction between a proton and a neutron could be written as

$$V(r) = J(r)P'_M P'_\sigma$$

in terms of the position and spin exchange operators P'_M and P'_σ for the pair. Thus for $J(r) > 0$ (Heisenberg's theory) the interaction energy of a proton–neutron pair would be positive if the wave function was symmetric for exchange of the position and spin co-ordinates and negative, if it was antisymmetric. The ground state of the nucleus would have as many antisymmetric proton–neutron pairs as possible. On the other hand if $J(r) < 0$ as suggested by Majorana, the ground state would have the greatest possible number of symmetric proton–neutron pairs. Therefore, the structure of the ground state of a nucleus would be completely different in the two theories.

Although this simple modification of Heisenberg's theory satisfied the saturation condition, it was still open to improvement. Majorana pointed out that this interaction could not explain the very large binding energy of the α-particle. The interaction $J(r)P'_M P'_\sigma$ with $J(r) < 0$ is attractive, if the wave function of a proton–neutron pair is completely symmetric for exchange of position and spin co-ordinates. In the deuteron the orbital wave function is almost certainly an S-state and therefore symmetric. It follows that the spin wave function is also symmetric and the spin of the ground state must be $S = 1$. (This agrees with measurements, but these experiments had not been made when Majorana wrote his paper.) Majorana pointed out that it was reasonable to assume that the ground state of the α-particle was completely symmetric in the position co-ordinates of all protons and neutrons. Then Heisenberg's interaction with $J(r) < 0$ gave a net attraction between neutrons and protons with parallel spins but no resultant force between those with opposite spins. Thus there were only two attractive bonds in the α-particle and its mean potential energy should be just twice that of the deuteron. On the other hand, the mean kinetic energy should be rather larger than twice that of the deuteron, because four nucleons are bound in a region of about the same size as the deuteron. So it seemed likely that the α-particle binding energy should be less than twice the binding energy of the deuteron, and it would be unstable for break-up into two deuterons. In any case the binding energy would be small, which was in disagreement with experimental results.

This contradiction led Majorana to consider an orbital exchange interaction

$$V(r) = -J(r)P'_M$$

with strength $J(r) > 0$ which is attractive in symmetric and repulsive in antisymmetric orbital states. In this interaction each proton in the α-particle interacts with both neutrons instead of with only one and vice versa. Thus there are four attractive bonds instead of only two. This is just the physical requirement for Wigner's theory to apply and the stability of the α-particle could be explained.

Majorana's theory was so satisfying aesthetically that it was accepted immediately. Heisenberg abandoned his force with a "repulsive core" and it was forgotten for nearly 20 years. Then, in 1952, some features of high energy proton–proton scattering could not be explained by purely attractive forces and Heisenberg's idea was revived. It is now generally agreed that two nucleons repel each other strongly when they are close together.

III
The Two-Body Problem

THE first theoretical investigations of nuclear structure described in Chap. II established important qualitative features of the nuclear forces. Calculations of the binding energy and size of nuclei made by Heisenberg and Fermi (1933) gave estimates of the strength and range of interaction between nucleons, but it was not possible to get reliable quantitative information because the methods of calculating nuclear energies were not accurate enough. Since then almost all our detailed knowledge of nuclear forces has come from studies of the two-nucleon problem, i.e. from properties of the deuteron and from proton–proton or neutron–proton scattering experiments.

Wigner wrote two fundamental papers in 1932 and 1933 on the deuteron and on the scattering of neutrons by protons. We discussed his investigation "On the Mass Defect of Helium" in Chap. II, but in this same paper he also made a detailed study of properties of the deuteron. In another paper he showed that the cross-section for scattering of neutrons by protons could be predicted from the deuteron binding energy. The two original papers are reproduced in Part 2 of this book.

3.1 Binding Energy and Size of the Deuteron

Wigner's paper on the mass defect of helium began with an investigation of the deuteron. He represented the neutron–proton interaction by a potential energy $V(r)$ depending only on r, the separation distance of the two particles. The force derived from this potential was attractive for all r and strong enough, so that the deuteron would have just one bound state with the observed binding energy. Wigner chose to use Eckart's (1930) potential for which the Schrödinger equation had simple analytic solutions. He concluded, however, that the results he derived were insensitive to the detailed shape of the potential function $V(r)$ provided that it had the same qualitative characteristics as Eckart's potential.

Wigner showed that the binding energy of the deuteron determined a relation between the range and the depth of $V(r)$. If the range of the potential is reduced the binding energy can be kept constant by increasing its depth. A graph in Fig. 2 of Wigner's paper (cf. Part 2, p. 175) illustrates the way the potential depth must vary, if the range is changed to keep the deuteron binding energy constant. The size of the deuteron on the other hand is rather insensitive to the range of the neutron–proton interaction and tends to a constant value determined entirely by the deuteron binding energy for a short range interaction.

If a number of potentials with different ranges and strengths adjusted to give the correct deuteron binding energy are used to calculate the He^4 binding energy, then, as a function of the interaction range, the He^4 binding energy increases as the range decreases and can be made arbitrarily large by an interaction with a sufficiently short range. Wigner derived this result (cf. Part 2, p. 170) and used it to show that the very large binding energy of He^4 (28·1 MeV) compared with the deuteron (2·2 MeV) could be explained by his model provided the neutron–proton interaction had a sufficiently short range. The latter must be rather less than the deuteron size.

Although Wigner conjectured that the low energy properties of the neutron–proton system were not sensitive to the shape of the interaction potential and supported his conjecture by an example, it was not until about 1947 that this idea was formulated clearly in the effective range theory of Schwinger (1947), Bethe (1949) and Blatt and Jackson (1949). The effective range theory is discussed in Section 3 of this chapter; here we are concerned with a few of Wigner's results that are independent of the shape of the potential function provided the deuteron is a "weakly bound" structure.

To define a weakly bound state we consider a family of attractive potentials $V(r) = -Uf(r)$. If the radial dependence $f(r)$ is kept fixed the strength of the potential is determined by the number U. If U is positive and small enough the neutron–proton system has no bound state. As U is increased a bound state will appear when U exceeds a critical value U_0. It is useful to measure the strength of the potential by a dimensionless parameter s defined by $s = U/U_0$. For $s < 1$ there is no bound state, for $s > 1$ there is at least one. We find one "weakly bound" state if U is just greater than U_0 or if s is only a little larger than unity.

We assume now that the potential $V(r)$ has a finite range, i.e. it becomes negligible for separations larger than some distance b. If there is only one bound state, it must be an S-state with zero orbital angular momentum. The wave function $u(r)$ satisfies Schrödinger's equation

$$-\frac{\hbar^2}{M}\frac{\mathrm{d}^2 u}{\mathrm{d}r^2} + V(r)u = -\varepsilon u \tag{3.1}$$

with boundary conditions

$$u(0) = 0 \quad \text{and } u(\infty) = 0$$

The quantity M is the nucleon mass (taken to be the same for both the neutron and the proton) and ε is the deuteron binding energy. For $r > b$, $V(r)$ is negligible and

$$u(r) = A \exp(-\gamma r) \tag{3.2}$$

where $\hbar^2\gamma^2/M = \varepsilon$ or $\gamma = (M\varepsilon)^{\frac{1}{2}}/\hbar$. The quantity γ^{-1} is a length and has the numerical value $\gamma^{-1} = 4{\cdot}33$ fm for the deuteron. If this state is weakly bound, i.e. if s is sufficiently close to unity, γ is very small in the sense that $\gamma b \ll 1$. (1 fm $= 10^{-13}$cm.)

The mean square radius of the bound state is given by

$$\langle r^2 \rangle = \int_0^\infty r^2 u(r)^2 \,\mathrm{d}r \Big/ \int_0^\infty u^2 \,\mathrm{d}r \tag{3.3}$$

If $\gamma b \ll 1$ then most of the contribution to the integrals in (3.3) comes from $r > b$, and the wave function can be replaced by its asymptotic form $u(r) = A \exp(-\gamma r)$ when evaluating them. The fractional error is of order γb. With this approximation we find that $\langle r^2 \rangle \approx \frac{1}{2}\gamma^{-2}$ which is much larger than b^2. In other words, the "size" of a weakly bound state is determined by its binding energy and is much larger

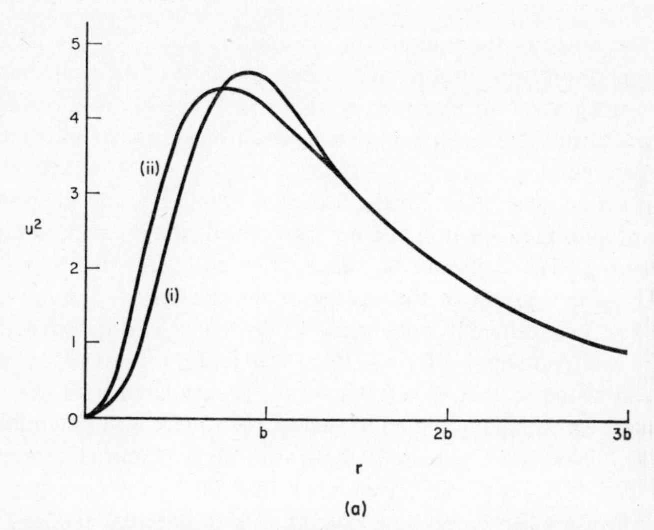

FIG. 2a. The square of the deuteron wave function $u^2(r)$ calculated with (i) a square well potential with range b, (ii) Eckart's potential with a range parameter $\rho = b/4$.

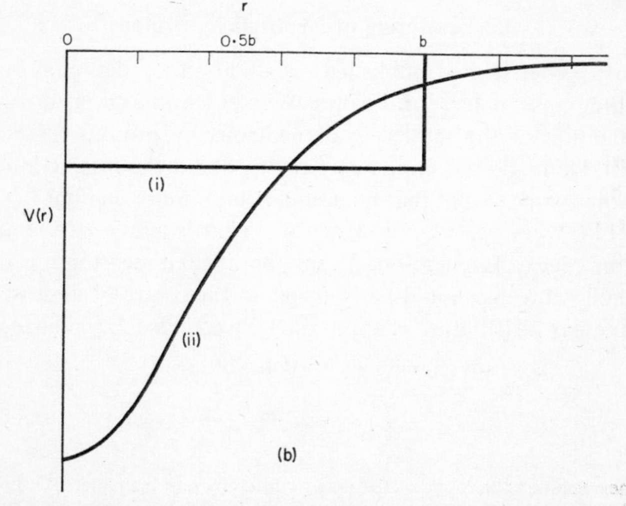

FIG. 2b. The potentials used to calculate the wave functions in Fig. 2a. (i) The square well potential, (ii) Eckart's potential.

than the range of the interaction binding the state. We can look at the same result from another angle: Because most of the contribution to the normalization integral $\int u^2 \, dr$ in (3·3) comes from $r > b$ the chances of finding the proton and neutron outside their interaction range are good.

The actual value of γb for the deuteron is about $\frac{1}{3}$, and the probability of finding the proton and neutron outside the interaction range is about $\frac{3}{4}$. The deuteron is weakly, but not very weakly bound. Figure 2a is a graph of the square of the deuteron wave function $u^2(r)$ for two different potentials. Curve (i) is calculated with a square well potential $V(r) = -V_0$, $r < b$ and $V(r) = 0$, $r > b$ with range adjusted so that $\gamma b = \frac{1}{3}$. Curve (ii) is calculated with Eckart's potential with range ρ related to that of the square well potential by $4\rho = b$. These two potentials have the same "intrinsic range". Blatt and Jackson (1950) have shown that this is the correct parameter to use when comparing potentials with different shapes. The two potentials are plotted in Fig. 2b and look quite different, but the wave functions calculated from them are almost identical.

3.2 Scattering of Neutrons by Protons

Shortly after he had published his study on the deuteron binding and the mass defect of helium Wigner produced a theoretical investigation of the scattering of neutrons by protons (cf. Part 2, p. 182). He made the same assumptions about the interaction as in his earlier work except that he used a square well potential $V(r) = 0$, $r > b$; $V(r) = -V_0$, $r < b$ instead of Eckart's potential.† For low neutron energy Wigner found that the angular distribution of the scattered neutrons should be isotropic in the centre-of-mass system. The angular distribution is approximately (cf. Part 2, p. 186, eq. 6a)

$$W(\theta) = (1 + \gamma b + 0{\cdot}4k^2 b^2 \cos\theta)$$
$$= \left(1 + \gamma b + 0{\cdot}2 \frac{E}{\varepsilon} (\gamma b)^2 \cos\theta\right) \qquad (3.4)$$

† Wigner denoted the range of the nuclear force by a in his paper. We prefer to use b, because a is the standard notation for the "scattering length" (cf. Chap. III, Section 3).

where $E = 2\hbar^2 k^2/M$ is the neutron energy as measured in the laboratory and ε is the deuteron binding energy. The helium study showed the deuteron was weakly bound so that $\gamma b \ll 1$. Thus the $\cos\theta$ term should be negligible unless $E \gg \varepsilon$. Alternatively, the experiments by Meitner and Philipps found no evidence of a $\cos\theta$ term in the angular distribution supporting the conclusion of the helium study that the interaction potential had a short range.

Wigner's approximate formula for the cross section can be written as

$$\sigma = 4\pi \frac{1+\gamma b}{\gamma^2 + k^2} \tag{3.5}$$

(This is obtained from Wigner's equation (7) by a change of variables expressing ε and E in terms of γ and k.)

If $\gamma b \ll 1$, the total cross section for low energy neutron scattering should be determined by γ, i.e. by the deuteron binding energy and not by the range of the forces. A measurement of the cross section could test Wigner's theory. If it turned out to be correct very accurate measurements might determine γb and hence the range of the forces.

The experimental value for the deuteron binding energy is $\varepsilon = 2{\cdot}2$ MeV and the limiting cross section for slow neutrons predicted from formula (3.5) (neglecting the term γb in the numerator) is

$$\sigma_{th} = 2{\cdot}3 \times 10^{-24}\ \text{cm}^2$$

The measured value $\sigma_0 = 20{\cdot}36 \times 10^{-24}\ \text{cm}^2$ deviates from the theoretical cross section by an order of magnitude. Early experiments were not very accurate, but the difference between the theoretical and experimental values was so large that it was noticed immediately. (Early measurements made by Dunning *et al.* (1935) and by Bjerge and Westcott (1935) found a value of about $30 \times 10^{-24}\text{cm}^2$.)

This striking failure of the theoretical prediction led to the discovery of a new physical effect. Wigner recognized the reason for the disagreement.† He had assumed that the nuclear force between

† Wigner's contribution was not published, but it is referred to in the review article of Bethe and Bacher, *Rev. Mod. Phys.* **8**, p. 117 (1937) and in the paper on proton–proton scattering of Breit, Condon and Present, *Phys. Rev.* **50**, p. 825 (1936).

a neutron and a proton was independent of the relative orientation of their spins. The experiments showed that this could not be true. The development of this idea is dealt with in Section 5 of Chap. III.

3.3 The Scattering Length and Effective Range

In this section we give a brief account of the modern theory of low energy neutron–proton scattering. Between 1947 and 1950 a systematic theoretical description of low energy scattering, called effective range theory, was developed by Schwinger (1947), Bethe (1949) and Blatt and Jackson (1949) and others. This theory shows that the scattering of one particle by another at low energies is characterized by two parameters, namely the *scattering length* and the *effective range*. Any potential that has its strength and range adjusted to reproduce the observed values of the scattering length and effective range will fit the experiments as well as any other. In this sense the low energy scattering data can fix the strength and range of the neutron–proton interaction potential but cannot determine its shape.

The scattering length was first introduced by Fermi (1934) and was used by him in his studies of the scattering of slow neutrons by nuclei.

In order to define the scattering length and effective range we first summarize some of the results of the partial wave method in scattering theory. A particle with mass m, energy E and wave number $k = (2mE)^{\frac{1}{2}}/\hbar$ moving parallel to the z-axis is influenced by a fixed scatterer. The particle interacts with the scatterer through a potential $V(r)$. The wave function $\psi(r)$ appropriate for a description of the scattering of the particle has the asymptotic form

$$\psi \sim \exp(ikz) + f(\theta) \exp(ikr)/r \quad \text{for large } r \tag{3.6}$$

In this expression the term $\exp(ikz)$ is the wave describing the incident particle and the term containing the scattering amplitude $f(\theta)$ is the scattered wave. The differential scattering cross section $d\sigma/d\Omega$ is related to the scattering amplitude by

$$\frac{d\sigma}{d\Omega} = |f(\theta)|^2 \tag{3.7}$$

If the potential is spherically symmetrical the scattering amplitude is determined by phase shifts δ_l in the various angular momentum states

$$f(\theta) = \frac{1}{k} \sum_l (2l+1) \sin \delta_l \exp(i\delta_l) P_l(\cos \theta) \tag{3.8}$$

The phase shift δ_l is defined by the asymptotic form of the radial wave function of the lth partial wave

$$u_l(r) \sim A_l \sin(kr + \delta_l - l\pi/2) \quad \text{as } r \to \infty \tag{3.9}$$

This in turn is found by solving the Schrödinger equation.

For small k we have $\delta_l \sim k^{2l+1}$, so that for very low energies only the term with $l = 0$ in (3.8) contributes to the scattering: only S-wave scattering is important. The radial wave equation satisfied by u_0 is

$$\frac{d^2 u_0}{dr^2} + k^2 u_0 = V u_0 \tag{3.10}$$

with boundary condition $u(0) = 0$. We assume that the potential $V(r)$ is negligible for $r > b$ and study the solutions for very small values of $k(kb \ll 1)$. In this calculation we use the fact that the shape of the wave function for $r < k^{-1}$ is almost independent of k. First we take the solution when $k = 0$. This is illustrated in Fig. 3. When

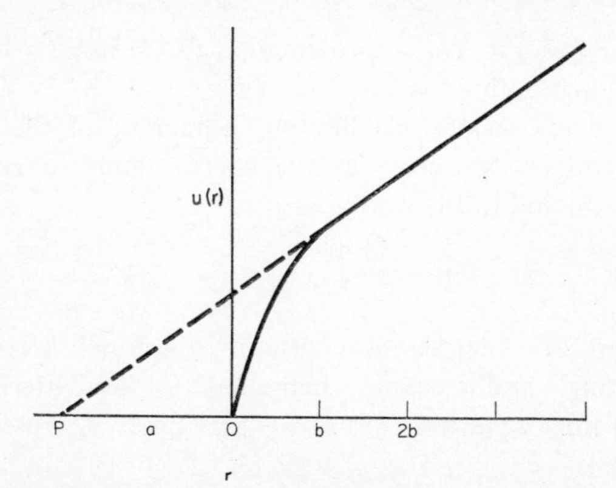

FIG. 3. Graph of a zero energy scattering wave function $u_s(r)$ to illustrate the definition of scattering length. The scattering length a is negative in this example.

$r > b$ the equation reduces to $d^2u_0/dr^2 = 0$ which has the solution

$$u_0(r) = c(1 - r/a) \quad \text{if } r > b \tag{3.11}$$

The constant a is determined by the intercept P of the linear part of the wave function with the r-axis. Its value depends on the wave function $u(r)$ in the region $0 < r < b$ and can be calculated for a given potential by solving the Schrödinger equation (3.10) with $k = 0$. This parameter is called the scattering length, and we shall show that it determines the limiting value of the cross section for small energies. The solution of equation (3.10) for $k \neq 0$ and $r > b$ is

$$u_0(r) = A_0 \sin(kr + \delta_0) = c' \sin(kr + \delta_0)/\sin \delta_0$$

where the normalization is chosen to facilitate comparison with equation (3.11). If $kb \ll 1$ then the sine function can be expanded for values of r in the range $b < r \ll k^{-1}$,

$$u_0(r) \approx c'(1 + kr \cot \delta_0) \tag{3.12}$$

Equations (3.11) and (3.12) are consistent if

$$k \cot \delta_0 \approx -\frac{1}{a} \tag{3.13}$$

when k is small. Thus

$$k^{-1} \tan \delta_0 \approx k^{-1} \sin \delta_0 \approx k^{-1} \delta_0 \approx -a \tag{3.14}$$

if k is small enough. The approximation (3.13) holds when $kb \ll 1$ and (3.14) when both $kb \ll 1$ and $ka \ll 1$.

Now we can consider the limiting behaviour of the scattering amplitude and cross section as the energy tends to zero. From equations (3.8) and (3.14) it follows that

$$f(\theta) \approx \frac{1}{k} \sin \delta_0 \approx -a \tag{3.15}$$

Thus for very low energies the scattering amplitude becomes independent of angle and is equal in magnitude to the scattering length. In the same limit equation (3.7) shows that the total cross section is

$$\sigma = 4\pi a^2 \tag{3.16}$$

Hence the limiting value of the total cross section measures the magnitude of the scattering length.

If the potential $V(r)$ has a bound state the bound state wave function is $u_0(r) \approx A \exp(-\gamma r)$ for $r > b$ with γ determined from the binding energy $\varepsilon(\gamma = (2m\varepsilon)^{\frac{1}{2}}/\hbar)$. In addition, if $\gamma b \ll 1$ the exponential function can be expanded as a power series in r and

$$u_0(r) \approx A(1-\gamma r) \tag{3.17}$$

for $b < r \ll \gamma^{-1}$. Comparing equations (3.17) and (3.11) we find that

$$a \approx \gamma^{-1} \tag{3.18}$$

This result is approximate and there are corrections of order γb. Equation (3.18) proves that $a > 0$, if the potential has a weakly bound state. If $a < 0$ one often says the potential has a *virtual state*.

It is interesting to relate equation (3.18) to Wigner's results for scattering from a square well potential (cf. Section 2.2). Comparing equations (3.16) and (3.5) as $E \to 0$ we find

$$\begin{aligned} a &= \gamma^{-1}(1+\gamma b)^{\frac{1}{2}} \\ &\approx \gamma^{-1}(1+\tfrac{1}{2}\gamma b) \end{aligned} \tag{3.19}$$

Hence Wigner's calculation gives the first correction to equation (3.18). Equation (3.19) shows that accurate experimental values of a and ε can determine the interaction range. The effective range theory derives results equivalent to (3.19) but without placing restrictions on the shape of the potential.

If the incident energy in a scattering experiment is not too high, the scattering amplitude is determined mainly by the S-wave phase shift δ_0. Schwinger proved† that

$$k \cot \delta_0 = -\frac{1}{a} + \tfrac{1}{2}\rho(E)k^2 \tag{3.20}$$

$$\simeq -\frac{1}{a} + \tfrac{1}{2}r_0 k^2 \tag{3.21}$$

where k is the wave number and a the scattering length. The quantity $\rho(E)$ has dimensions of length and is usually such a slowly varying function of the energy E that it can be replaced by a constant r_0 called the effective range over a wide energy range. Equation (3.21)

† We do not prove the effective range equations because derivations appear in many text-books. Most derivations use a method due to Bethe (1949).

is a good approximation for energies up to about 20 MeV in neutron–proton scattering. Measurements of the scattering at several energies can determine a and r_0.

The effective range formula can be extended to apply to a bound state and gives a relation between a, r_0 and γ

$$\gamma \simeq \frac{1}{a} + \tfrac{1}{2} r_0 \gamma^2 \tag{3.22}$$

This equation is the generalization of equation (3.19) derived from Wigner's paper for a square well potential. Comparison of equation (3.22) with (3.19) for small values of r_0 shows that the effective range r_0 of a square well potential is almost equal to its actual range b.

The effective range theory is very general and holds even if the neutron–proton interaction cannot be represented by a potential at all. The only restriction is that the interaction should have a reasonably short range. The theory does not apply if Coulomb interactions are present unless it is modified.

3.4 Spin-Dependence of the Neutron–Proton Force

In Section 2 of this chapter we saw that Wigner's prediction for the scattering cross section of slow neutrons by protons disagreed with the experimental value by an order of magnitude suggesting that the neutron–proton forces are spin dependent. Because the proton and neutron have each a spin quantum number $\frac{1}{2}$ their total spin can be either $S=0$ (singlet states) or $S=1$ (triplet states). In low energy scattering only S-waves are important, so the orbital angular momentum L is zero and the spin S is equal to the total angular momentum J. The total angular momentum is conserved, hence the spin is also conserved, and the scattering in singlet and triplet states can be treated independently by the effective range theory. The low energy scattering is determined by singlet and triplet scattering lengths a_s and a_t and effective ranges r_{0s} and r_{0t} respectively. If both neutrons and protons are unpolarized there are four equally probable spin states. Three of them are triplet states and one is a singlet state.

Hence in scattering the triplet and singlet states occur with probabilities $\frac{3}{4}$ and $\frac{1}{4}$ respectively. The total zero energy cross section is therefore

$$\begin{aligned}\sigma_0 &= \tfrac{3}{4}\sigma_t + \tfrac{1}{4}\sigma_s \\ &= \pi(3a_t^2 + a_s^2) \qquad (3.23)\end{aligned}$$

In 1934 the spin of the deuteron ground state was measured by Murphy (1934) and found to be $I = 1$. Thus the deuteron binding energy determines the triplet scattering length approximately through equation (3.18) and the triplet cross section σ_t from equation (3.16). We estimated σ_t in Section 2.2 and found it was much less than the experiment cross section. This result is consistent with equation (3.23) provided $\sigma_s \gg \sigma_t$ or $|a_s| \gg |a_t|$. We can even estimate the magnitude of a_s from the measured cross section σ_0 but cannot fix its sign. If $a_s > 0$ there would be a bound singlet state of the deuteron with a binding energy of about 75 keV. If $a_s < 0$ there would be a virtual state. In either case we conclude that the neutron–proton interaction is weaker in the singlet than in the triplet state. Wigner recognized this in 1935. But there was still a question to answer: Is the singlet state of the deuteron bound or virtual?

The first solution to this problem was suggested by Fermi (1936). If a thermal neutron moves through a substance containing hydrogen it may be captured by a hydrogen nucleus emitting a γ-ray to form a deuteron. Fermi and Amaldi had measured the mean life of a neutron against capture by hydrogen in paraffin and found it to be $\tau = 1{\cdot}7 \times 10^{-4}$ sec. The most probable capture process involves emission of magnetic dipole γ-radiation accompanied by a transition of the neutron from a singlet $^1S_0$ scattering state to the triplet $^3S_1$ ground state of the deuteron. Fermi calculated the cross section for this process and found that it contained a factor M^2 which was the square of the overlap between the initial scattering state wave function $u_s(r)$ and the deuteron ground state $u_0(r)$

$$M^2 = \left|\int u_s(r) u_0(r)\, \mathrm{d}r\right|^2$$

The two possible scattering wave functions ($a_s > 0$ or $a_s < 0$) are sketched in Fig. 4 with the ground state wave function. The overlap

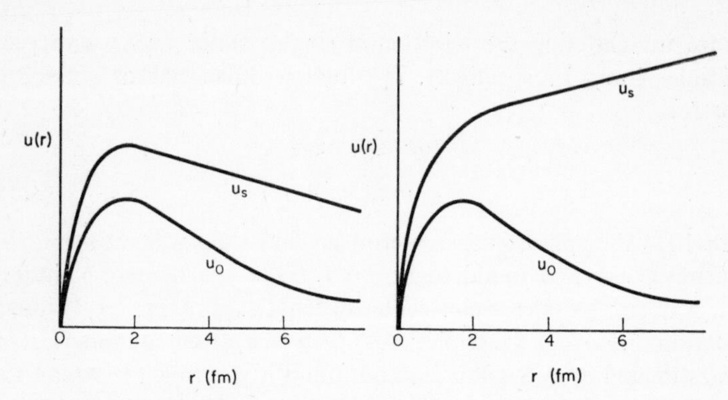

FIG. 4. Illustration of the deuteron ground state wave function $u_0(r)$ and the singlet neutron–proton scattering wave function $u_s(r)$. (a) If the singlet scattering length $a_s > 0$, (b) If $a_s < 0$.

is clearly greater when $a_s < 0$, or when the $^1S_0$ state is virtual. If most of the contribution to the overlap comes from values of r outside the range of the interaction, M^2 is proportional to $(a_s - a_t)^2$. Fermi calculated the neutron capture life time in paraffin from the data available and found that

$$\tau = 6{\cdot}5 \times 10^{-4} \text{ sec if } a_s > 0 \quad (\text{bound } ^1S_0 \text{ state})$$

and

$$\tau = 2{\cdot}6 \times 10^{-4} \text{ sec if } a_s < 0 \quad (\text{virtual } ^1S_0 \text{ state})$$

Comparing these results with the experimental value he concluded that the $^1S_0$ state was virtual or $a_s < 0$.

Another, completely different method involving coherent scattering of thermal neutrons from hydrogen molecules was suggested by Schwinger and Teller (1937). This method was more sensitive than Fermi's and at the same time it made it possible to determine both a_s and a_t from scattering experiments without having to use the deuteron binding energy to find a_t.

When a neutron is scattered by a hydrogen molecule the components of the neutron wave scattered by the individual atoms in the molecule interfere. This effect becomes very important when the neutron wave length is comparable with or larger than the size of the molecule, and also, when the neutron energy is smaller than the spacing between molecular energy levels, so that inelastic scattering does not occur.

If the neutron–proton scattering amplitude is spin dependent the coherent scattering is influenced by the relative orientation of the proton spins in the hydrogen molecule. Hence, the scattering by ortho-hydrogen (total proton spin $I = 1$) and para-hydrogen ($I = 0$) should be different. The spacing between the two lowest rotational levels of the hydrogen molecule is 0·015 eV, and the average kinetic energy of a thermal neutron at liquid air temperature is $3kT/2 = 0{\cdot}012$ eV. At this temperature a neutron wave length ($\lambda = 2{\cdot}8 \times 10^{-8}$ cm) is rather larger than the molecular diameter ($d = 0{\cdot}74 \times 10^{-8}$ cm), so the conditions required for coherent scattering are satisfied. The first experiments were made in 1937 with thermal neutrons at liquid air temperatures, while the most accurate measurements used thermal neutrons with a temperature of 20° K.

The coherent scattering of neutrons by hydrogen molecules can be calculated simply in the limit of very low energies when the neutron wave length is much larger than the molecular diameter. We make use of the result (Section 2.3) that the scattering amplitude is equal to the scattering length under these circumstances. The general amplitude for scattering of a neutron by a proton can be written as

$$\begin{aligned} a &= a_s + \tfrac{1}{2}(a_t - a_s)S(S+1) \\ &= a_s + \tfrac{1}{2}(a_t - a_s)\mathbf{S}^2 \end{aligned} \tag{3.24}$$

This amplitude reduces to a_s for singlet scattering ($S = 0$) and a_t for triplet scattering ($S = 1$), but it can also be used when the scattering is neither singlet nor triplet but a mixture of both. The total spin $\mathbf{S} = \mathbf{s}_n + \mathbf{s}_p$ is the vector sum of the neutron and proton spins. We have

$$\begin{aligned} \mathbf{S}^2 &= \mathbf{s}_n^2 + \mathbf{s}_p^2 + 2\mathbf{s}_n \cdot \mathbf{s}_p \\ &= 2\mathbf{s}_n \cdot \mathbf{s}_p + \tfrac{3}{2} \end{aligned}$$

where we have put $\mathbf{s}_n^2 = s_n(s_n + 1) = \frac{3}{4}$, because the neutron has spin $s_n = \frac{1}{2}$ and similarly for $\mathbf{s}_p^2$. If this result is substituted into equation (3.24) we get

$$a = \tfrac{1}{4}(3a_t + a_s) + (a_t - a_s)\mathbf{s}_n \cdot \mathbf{s}_p \tag{3.25}$$

The amplitude for scattering of a neutron by a hydrogen molecule is the sum of the amplitudes for scattering from the individual nuclei in the molecule

$$\begin{aligned} a_{\text{mol}} &= a_1 + a_2 \\ &= \tfrac{1}{2}(3a_t + a_s) + (a_t - a_s)\mathbf{s}_{\text{n}} . \mathbf{I} \end{aligned} \tag{3.26}$$

where $\mathbf{I} = \mathbf{s}_{\text{p1}} + \mathbf{s}_{\text{p2}}$ is the total spin operator of the molecule. To calculate the scattering from para-hydrogen we note that the total spin of the protons in the molecule is $\mathbf{I} = 0$, so the second term of equation (3.25) does not contribute and

$$a_{\text{para}} = f = \tfrac{1}{2}(3a_t + a_s) \tag{3.27}$$

The para-hydrogen cross section is just $4\pi f^2$. The quantity f is called the coherent scattering length. Our calculation is valid only if the neutron wave length is very long compared with the size of the hydrogen molecule. In practice this condition is never realized, and the formulae for the scattering amplitude must be modified in order to extract accurate values of the coherent scattering length f from the experimental data. The original paper of Schwinger and Teller (1937) contained the most important corrections and for this reason their calculation is much more complicated than the one we have given here.

The magnitude of f is very sensitive to the relative sign of a_s and a_t. Thus even a very rough measurement of the scattering cross section of neutrons by para-hydrogen at liquid air temperature is sufficient to determine the sign of a_s. The first experiment by Halpern (1937) showed that the singlet state of the deuteron was virtual in agreement with Fermi's result.

Since 1937 experimental techniques have improved and accurate measurements of f have been made using neutron scattering from para-hydrogen (Squires 1953). Other methods have also been developed. In the most accurate of these the refractive index of substances containing hydrogen is found by measuring the critical angle for external reflection of neutrons (Hamermesh 1950, Hughes *et al.* 1950). If a wave passes through a medium containing a large number of scattering centres, the macroscopic refractive index n of the medium is related to the amplitude for scattering from the individual centres by

$$n^2 - 1 = -\lambda^2 Na/\pi$$

where λ is the wavelength, N the number of scattering centres per unit volume and a the scattering amplitude. If the medium contains several different kinds of centres, then a is the average of the various scattering amplitudes. For liquid hydrogen this average amplitude would be $a = \frac{1}{4}(3a_t + a_s) = f/2$, because a triplet state is three times more probable than a singlet state. The contribution of the hydrogen to the refractive index of a substance is always determined by the coherent scattering length f, and refractive index measurements can find this quantity.

Recent experimental values of f and σ_0 are given in Table 1 and derived values of a_s and a_t calculated from equations (3.27) and

TABLE 1

Experimental Low Energy n-p *Parameters*

$$\varepsilon = 2{\cdot}2245 \pm 0{\cdot}0002 \text{ MeV}$$
$$\gamma = 0{\cdot}23169 \pm 0{\cdot}00001 \text{ fm}^{-1}$$
$$\sigma_0 = 20{\cdot}36 \pm 0{\cdot}05 \times 10^{-24} \text{ cm}^2$$
$$f = -3{\cdot}741 \pm 0{\cdot}011 \text{ fm}$$

TABLE 2

Low Energy Parameters in the Two-Nucleon Interaction

	Scattering length (fm)	Effective range (fm)
Singlet p–p	$-7{\cdot}778 \pm 0{\cdot}007$	$2{\cdot}714 \pm 0{\cdot}011$
Singlet n–p	$-23{\cdot}680 \pm 0{\cdot}028$	$2{\cdot}46 \pm 0{\cdot}12$
Triplet n–p	$5{\cdot}399 \pm 0{\cdot}011$	$1{\cdot}732 \pm 0{\cdot}012$

(3.23) in Table 2. Notice a_t and γ^{-1} (4·32 fm) are not exactly equal. The difference is due to the finite range of the neutron–proton force, and the triplet effective range r_{0t} can be calculated from it by using equation (3.22). It is harder to measure the singlet effective range r_{0s}.

A value can be found by measuring the total cross section at several energies and by analysing the results using the effective range theory (equation 3.21). The errors are much larger than with other parameters. (For a detailed summary of the experimental data see Wilson (1963).)

3.5 Proton–Proton Scattering

The first experimental investigations of proton–proton scattering were made in 1935 and 1936 using the cyclotron developed by Lawrence and Livingston at Berkeley and the Van de Graaff of Tuve and his collaborators at the Department of Terrestrial Magnetism in Washington. These machines were operating in 1932 (cf. Part 2, p. 131) and by 1935 they were sufficiently developed to make a serious study of proton–proton scattering possible.

If the Coulomb force is the only interaction between protons, the angular distribution of the scattered protons is similar to classical Rutherford scattering with some quantum mechanical corrections due to the identity of the particles. The corrections had been calculated by Mott in 1930 and were applied to (α, α) scattering in 1932. White working at Berkeley in 1935 found evidence that the experimental angular distribution deviated from Mott's theoretical prediction. In his experiment scattered protons were observed in a cloud chamber. It was difficult to get good statistical accuracy as the number of events observed at large angles was rather small. Although the deviation from Mott's theory seemed significant, quantitative conclusions could not be drawn from the results.

At about the same time, Tuve, Heydenberg and Hafstad (1936) were working with the Van de Graaff in Washington using an electronic method for counting scattered protons. Large numbers of scattered particles were observed and their data were free from the statistical fluctuations present in White's measurements due to an insufficiency of such particles. The energy calibration of the proton beam used in their experiments was also rather accurate. In 1936 they published their results giving angular distributions for several proton energies. They confirmed the deviations from Mott's law found by White.

The results were analysed by Breit, Condon and Present (1936) who showed that the deviations from Mott's law could be explained, if there was a short range attractive nuclear interaction between protons in addition to the Coulomb forces. The experimental results could not fix both the strength and the range of the interaction. However, if the ranges of the neutron–proton and proton–proton interactions were taken to be equal their strengths were also approximately the same. (This fact led to the hypothesis of *charge independence* for nuclear forces. We shall discuss this in Chap. IV.)

The experiments could not determine the range of the proton–proton force for the following reason: Deviations from Coulomb scattering at low energies are only important when the relative orbital angular momentum of the protons is zero. A state with orbital angular momentum zero must be a singlet state ($^1S_0$), because of the Pauli exclusion principle. Thus, the deviations are determined by the $^1S_0$ phase shift. A measurement of proton–proton scattering at one energy can fix only one parameter, namely this phase shift. To find the strength and range of the interaction we need to determine two parameters; so the $^1S_0$ phase shift must be measured at two different energies. To get accurate values these energies have to be widely separated. The separation of the maximum energy (900 keV) and minimum energy (680 keV) in the experiments by Tuve, Heydenberg and Hafstad was not sufficient to give an accurate determination of the range and could only fix an upper limit.

Breit has been associated with the investigations of proton–proton scattering right from the beginning up to the present time. In 1926 he began investigations on accelerator design. These were continued by Tuve and his collaborators, and in 1932 the first electrostatic generator of the Van de Graaff type was used in nuclear studies. This accelerator was built to study proton–proton scattering, and we discussed the first results obtained with it in the previous paragraph. Breit had moved to the University of Wisconsin, but he was still connected with the project and analysed the results with Condon and Present. In 1936 another Van de Graaff was built at the University of Wisconsin. It could accelerate protons up to

2·5 MeV and by 1939 measurements of the proton–proton scattering had been extended up to that energy (Herb, Kerst, Parkinson and Plain, 1939). These results were analysed by Breit, Thaxton and Eisenbud (1939). Assuming that the nuclear interaction between protons could be represented by a square well potential they found a range of about $2{\cdot}8 \times 10^{-13}$ cm. Breit and his colleagues at Yale are still active in interpreting nucleon–nucleon scattering data, cf. Ch. V (Breit 1960, 62).

There is an effective range expansion for the $^1S_0$ phase shift in proton–proton scattering which is similar to the neutron–proton effective range formula, but it is more complicated because of the long range Coulomb repulsion. Low energy scattering data again determine a scattering length and an effective range. Recent values are given in Table 2 (cf. p. 41).

3.6 The Tensor Force

Investigations of nuclear structure described in Chap. II assumed that the nuclear force could be represented by a combination of exchange potentials with strengths depending only on the separation of a pair of interacting nucleons. The force derived from one of these potentials is called a central force: Its line of action passes through the pair of interacting particles although its magnitude may depend on the relative orientation of their spins. Central forces were also assumed in the discussion of low energy nucleon–nucleon scattering. These forces conserve spin angular momentum S and orbital angular momentum L separately as well as the total angular momentum J and parity π.

The quadrupole moment of a charge distribution $\rho(r)$ is defined by

$$Q = \int (3z^2 - r^2)\rho(r)\,d\mathbf{r}$$

If the charge distribution is spherically symmetrical then $Q = 0$. With central forces the deuteron would have orbital and spin quantum numbers $L = 0$ and $S = 1$. A state with $L = 0$ is spherically symmetrical, hence the deuteron should have zero quadrupole moment. The quadrupole moment was measured in 1939 by Kellogg,

Rabi, Ramsey and Zacharias (cf. Part 2, p. 189) who found that its value was not zero. Recent measurements (Auffray 1961) show that

$$Q = 2{\cdot}28 \times 10^{-27} \text{ cm}^2$$

These experiments refute the hypothesis of a pure central force and show that the interaction between a neutron and a proton must be of a more complex character.

Kellogg and his collaborators published their first results in a *Physical Review* letter in January 1939 and later in a paper (Kellogg 1940). We reproduce the original letter in Part 2 of this book and in the following paragraphs introduce the theory underlying the experiments.

The quadrupole moment was detected in a molecular beam resonance experiment using first hydrogen and then deuterium molecules. In the experiment a beam of hydrogen or deuterium molecules passing through a strong homogenous magnetic field was subject to a radio frequency field. The strength of the homogenous field was varied and resonances in the beam intensity at the detector were observed. The resonance patterns are shown in Figs. 1 and 2 of the original letter (cf. Part 2, p. 190). If the deuteron had zero quadrupole moment the result of the deuterium experiment could be predicted from the the measurements on the hydrogen molecules. The observed resonance pattern in deuterium was quite different from the predicted one, hence the deuteron quadrupole moment could not be zero (cf. Fig. 6b).

A molecule in a homogenous field can exist in one of several discrete energy states depending on the orientation of the molecular angular momentum relative to the direction of the field. If the energy difference ΔE between two of these states is related to the frequency ν of the r.f. field by Bohr's relation $\Delta E = h\nu$ then the transition between these states can occur. The resonances can be observed by varying either the homogenous field strength or the frequency of the r.f. field. The resonance patterns determine the energy states of the molecule in the magnetic field.

The number of resonance lines and their spacing depend on details of the molecular energy states: The H_2 molecule can exist in various excited rotational states which fall into two groups depending on the total spin I of the protons in the molecule. Each proton has spin $\frac{1}{2}$, so that the total spin of the two hydrogen nuclei can be either $I=0$ (para-hydrogen) or $I=1$ (ortho-hydrogen). If $I=0$ the rotational angular momentum† J must be an even integer ($J=0, 2, 4, \ldots$). If $I=1$, J must be odd. The restrictions on J are due to the Pauli exclusion principle. In the hydrogen molecular beam experiment measurements were made on the first rotational state of the molecule with $I=1$ and $J=1$. There are three possible orientations of the nuclear spin corresponding to magnetic quantum numbers $M_I=1$, 0, -1 and three for the rotational angular momentum with $M_J=1$, 0, -1, i.e. nine states all together. The energies of these nine states split into three groups of three in a homogenous magnetic field H, and in the absence of molecular perturbations the splitting would be as shown in Fig. 5 with equal spacing $\mu_R H$ within each group and equal spacing $2\mu_P H$ between the groups. (μ_R is the rotational magnetic moment of the molecule and μ_P is the proton moment; $\mu_R < \mu_P$). Transitions involving a change of nuclear spin direction are restricted by a selection rule $\Delta M_I = \pm 1$ and $\Delta M_J = 0$. The six possible transitions are indicated in Fig. 5. Each transition has a frequency $\nu = 2\mu_P H/\boldsymbol{h}$. Transitions involving a change in direction of the rotational angular momentum have a much lower frequency and were not studied in this experiment. If there are molecular perturbations the nine energy states are shifted a little and the resonance frequencies change so that there are six resonance lines grouped about this frequency instead of one.

Two perturbations split the resonance line in the H_2 molecule. One is a spin-orbit coupling between the nuclear spins and the rotation of the molecule. It can be represented by an energy $\mu_P \overline{H}\, \mathbf{J}.\mathbf{I}$, where $\overline{H}$ measures the coupling strength. The other is the interaction between the magnetic moments of the two nuclei and has a potential energy

$$V(r) = \boldsymbol{\mu}_1.\boldsymbol{\mu}_2/r^3 - 3(\boldsymbol{\mu}_1.\boldsymbol{r})(\boldsymbol{\mu}_2.\boldsymbol{r})/r^5 \qquad (3.28)$$

† In this discussion we use the notation of Kellogg *et al.* and J denotes the rotational angular momentum of a molecule. In other parts of our book J stands for the nuclear spin.

where r is the distance between the nuclei in the molecule. To the first order in perturbation theory the energy of the level (M_I, M_J) is given by the expression

$$E(M_I, M_J) = 2\mu_P M_I H + \mu_R M_J H + AM_I M_J + B(3M_I^2 - 2)(3M_J^2 - 2) \tag{3.29}$$

The first two terms are the zero order energies shown in Fig. 5 and the last two terms give the effect of the molecular perturbations

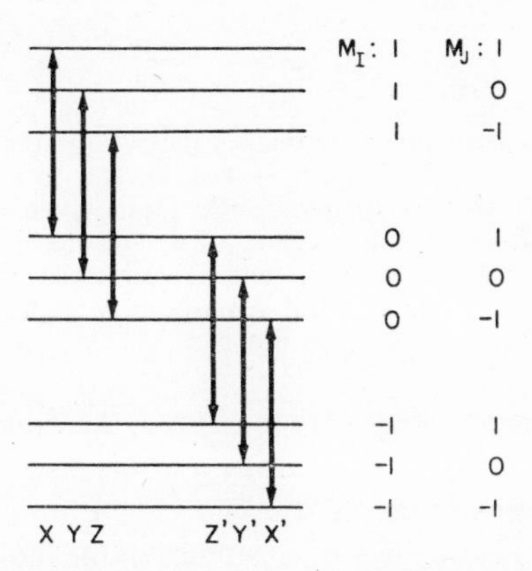

Fig. 5. The splitting of an $I = 1$, $J = 1$ molecular state in a negative field. The six transitions allowed by the selection rule $\Delta M_I = 1$, $\Delta M_J = 0$ are indicated by arrows and labelled by X, Y, Z, X', Y', Z'.

($A = \mu_P \overline{H}$, and $B = 2\mu_P^2/5r^3$). Equation (3.29) gives the most general first order shift which can occur in a molecular state with $I = 1$ and $J = 1$. Other physical perturbations can only modify the values of the coefficients A and B. There are small second order correction terms which have to be taken into account in a quantitative analysis of the experimental results (Kellogg *et al.* 1940). The energies of the transitions allowed by the selection rule are

$$E(1, M_J) - E(0, M_J) = 2\mu_P H + AM_J + 3B(3M_J^2 - 2)$$
$$E(0, M_J) - E(-1, M_J) = 2\mu_P H + AM_J - 3B(3M_J^2 - 2) \tag{3.30}$$

Fig. 6a shows the resonance pattern expected if the coefficient A is large compared with B and Fig. 6b the pattern if B is large compared

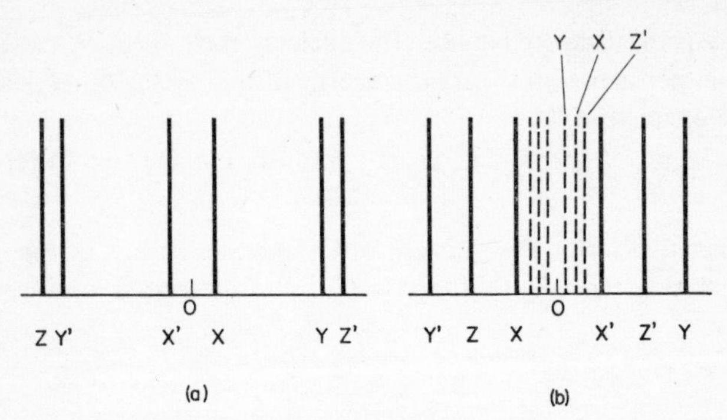

FIG. 6a. Diagram of the resonance pattern for the H_2 molecule ($A \simeq -4B$).

FIG. 6b. The resonance pattern for the D_2 molecule. The solid lines illustrate the observed pattern ($A \simeq -B$) and the dotted lines the pattern expected if the deuteron quadrupole moment were zero. The resonance lines X, Y, Z, X', Y', Z' correspond to transitions indicated in Fig. 5.

with A. If the resonance fields are measured, the values of μ_P, $\overline{H}$ and r can be extracted from them. Kellogg and his co-workers found μ_P, $\overline{H}$ and r from their results on H_2. The separation r of the nuclei in the molecule was also known from studies of the molecular spectrum of hydrogen and could be compared with the value from the molecular beam experiment. The two values were in excellent agreement checking the consistency of the method.

Deuterons have spin 1 and obey Bose–Einstein statistics; thus the total nuclear spin $I=0$ and $I=2$ can occur with the J even rotational states and $I=1$ with the odd J states. In the deuteron experiment the states $I=2$, $J=0$ and $I=1$, $J=1$ contributed to the observed resonance spectrum. There are no perturbation effects in the $I=2$, $J=0$ state and it produces one strong resonance line with frequency $\mu_D H/h$. This is the central line of Fig. 2 of Kellogg *et al.* (cf. Part 2, p. 190). The six small peaks in Fig. 2 of Kellogg's paper come from the second state $I=1, J=1$ which has the same angular momentum quantum numbers as the state investigated in hydrogen. The molecular perturbations of the energy levels should be represented by an equation similar to (3.29) with values of A and B that can be calculated from μ_D and from the $\overline{H}$ and r obtained from the hydrogen experiment, if the only perturbations are due to the spin–orbit and spin–spin interaction. The observed spectrum could be fitted, but the value of B was 30 times larger than the one predicted by the spin–spin interaction proving that some other perturbation must be important. A deuteron quadrupole moment $Q = 2 \times 10^{-27}$ cm^2 could explain the experimental results.

The deuteron was known to have a total angular momentum $J = 1$ and was thought to be a $^3S_1$ state, but the discovery of the non-zero quadrupole moment proved that this could not be so. In Russell-Saunders coupling there are three other states with $J = 1$, namely $^3P_1$, $^1P_1$ and $^3D_1$. Of these the $^3D_1$ state has the same parity as the $^3S_1$. If there is some nuclear interaction which breaks down Russell-Saunders coupling then the ground state of the deuteron could be some linear combination of the $^3S_1$ and $^3D_1$ states

$$\psi = \phi(^3S_1) + \alpha\phi(^3D_1) \tag{3.31}$$

and its quadrupole moment would be related to the amplitude α of the $^3D_1$ admixture. Only a small admixture ($\alpha \simeq 0{\cdot}3$) is needed to get the observed quadrupole moment.

If this explanation of the quadrupole moment is correct then the nuclear force between a neutron and a proton must contain some non-central component which mixes states with the same J but different L and S. A force with this property, called the tensor force, had been predicted by Yukawa (1938) and Kemmer (1938) from a modification of Yukawa's meson theory of nuclear forces. The potential of the tensor force is

$$\begin{aligned} V(r) &= V_T(r) S_{12} \\ &= V_T(r)\{3(\boldsymbol{\sigma}_1 . \hat{\mathbf{r}})(\boldsymbol{\sigma}_2 . \hat{\mathbf{r}}) - \boldsymbol{\sigma}_1 . \boldsymbol{\sigma}_2\} \end{aligned} \tag{3.32}$$

where $\boldsymbol{\sigma}_1$ and $\boldsymbol{\sigma}_2$ are the Pauli spin matrices for the two nucleons, $\hat{\mathbf{r}}$ is a unit vector parallel to the line joining them and $V_T(r)$ is some scalar function of their separation.

The interaction between two dipoles is represented by a potential with the same structure as the potential of the tensor interaction (cf. equation (3.28)), so the respective forces must have similar qualitative features. One dipole exerts a force on another which is equal in magnitude but opposite in direction to the force on the first due to the second. These forces do not have the same line of action, however, and their direction is not parallel to the line joining the dipoles. Hence, these two forces have a non-zero moment, and it seems at first sight that the total moment of the internal forces between the dipoles is not zero. There are couples acting on the dipoles, however, which exactly balance the moment of the forces between them. The tensor force between nucleons has the same characteristics. It is non-central in the same sense as the force between dipoles and, because the forces between the nucleons have a non-zero moment which is exactly compensated for by couples acting on the spins, there can be a transfer of angular momentum from the orbital motion to the spins and vice versa. For this reason only the total angular momentum is conserved and not the orbital and spin angular momenta separately. States with the same J but different L and S can be mixed, and there is a breakdown of Russell-Saunders coupling.

The tensor force has another important property which has a quantum origin and has no analogue in classical mechanics. If $\mathbf{S} = \frac{1}{2}(\boldsymbol{\sigma}_1 + \boldsymbol{\sigma}_2)$ is the total spin operator of the two nucleons, the tensor force operator S_{12} (equation (3.32)) can be expressed as a function of $\mathbf{S}$

$$S_{12} = \tfrac{1}{2}(3(\mathbf{S}.\hat{\mathbf{r}})^2 - \mathbf{S}^2) \tag{3.33}$$

In deriving this relation one has to use special properties of the Pauli spin matrices, i.e. $\boldsymbol{\sigma}^2 = 3$ and $(\sigma_{\mathbf{n}})^2 = 1$ ($\sigma_{\mathbf{n}}$ is the component of $\boldsymbol{\sigma}$ resolved along any unit vector $\mathbf{n}$) and it has no classical counterpart. One consequence of equation (3.33) is that the tensor force does not act in singlet states ($S = 0$) of two particles. It therefore contributes nothing to the singlet S-wave proton–proton or proton–neutron scattering, and the effective range theory described in the first part of this chapter requires no modification. In a triplet state this situation is more complicated. D-waves do not contribute to the low energy scattering, and the data still determine a scattering length and an effective range, but these parameters can be fitted by potentials which contain different mixtures of central and tensor forces. In other words, the scattering data give no information about the relative strength of the central and tensor parts. For potentials of a given shape the deuteron quadrupole moment fixes the relative strengths. The strength of the central part of the potential needed to fit the data is considerably less than that required if tensor forces are neglected. To produce the same deuteron binding energy in the presence of tensor forces, the central part need not be so strong as would be necessary otherwise. Although the admixture of D-state wave function is only a few per cent the tensor force makes a large contribution to the deuteron binding energy.

The situation is different for the α-particle. Because the spins of all four nucleons are paired off and the α-particle has a closed shell structure the tensor force makes only a small contribution to its binding energy. If the α-particle binding energy is calculated in two cases: (a) With pure central forces fitted to scattering data, and (b) with central plus tensor forces fitted to scattering data, the binding energy turns out to be less in case (b); i.e. when tensor forces are included, because the tensor forces do not contribute much to the binding energy and the central forces are less strong in case (b) compared with case (a). Thus old calculations neglecting tensor forces gave binding energies much larger than the experimental value, whereas modern calculations including tensor forces tend to give binding energies which are too small (cf. Section 2.3).

Although the tensor force is believed to be quite strong its influence on the properties of nuclei is not well understood. This is partly because its effects are mixed up with those of the spin–orbit force as will be discussed in the next section.

3.7 The Spin–Orbit Force

During the decade following the discovery of the neutron hopes of nuclear forces showing some simple features gradually faded. In 1933 Majorana set himself the problem of finding the simplest interaction law that would explain the saturation phenomenon. Eight years later, Wigner and Eisenbud (1941) wrote a paper giving the most general potential consistent with the laws of conservation of angular momentum and parity and the requirement of time reversal invariance. The most general potential which can act between identical nucleons and which does not involve the momentum to a power higher than the first is a linear combination of four basic types:

$$V(r) = V_1(r)+V_2(r)\boldsymbol{\sigma}_1 . \boldsymbol{\sigma}_2+V_T(r)S_{12}+V_{so}(r)\mathbf{L}.\mathbf{S} \qquad (3.34)$$

i.e. a combination of a Wigner force, a spin dependent central force, a tensor force and a spin–orbit force represented by the last term in equation (3.34). (In equation (3.34) $\mathbf{L} = \mathbf{r}\times\mathbf{p}$, where $\mathbf{r}$ and $\mathbf{p}$ are the relative coordinates and momenta of the two nucleons and

$$\mathbf{S} = \tfrac{1}{2}(\sigma_1 + \sigma_2)$$

is their total spin.) In the neutron–proton interaction each of these four types could also be combined with a Majorana exchange operator giving eight possible types altogether.

The first indication of a spin–orbit component in the nucleon interaction came from the (j,j) coupling shell model of Mayer (1950) and Jensen (1950). The shell structure of nuclei showed that, if the interaction of a single nucleon with a complex nucleus could be represented by an average potential, then this average potential must have a spin–orbit component. This could arise from a spin–orbit term in the fundamental nucleon–nucleon interaction. Unfortunately, the tensor force can also give a spin–orbit coupling in the nucleon–nucleus potential. For this reason, it has not proved possible to argue back from the shell model potential to estimate the strength of the nucleon–nucleon spin–orbit interaction. The most unambiguous evidence for a spin–orbit force comes from the analysis of high-energy proton–proton scattering.

IV

Charge Symmetry and Charge Independence

In his first paper on the proton–neutron model of the nucleus Heisenberg (cf. Part 2, p. 144) recognized that the forces acting between pairs of neutrons must be almost equal to those between pairs of protons. The argument he used depended on the observed variation of the charge mass ratio of stable elements through the periodic table. He concluded that this variation was due to the electro-static repulsion between protons. If it was not for this the ratio N/Z would be almost constant for all elements, and the value of this constant would depend on the relative strength of neutron–neutron and proton–proton interactions. The empirical data showed that it would be nearly unity, equal to its value in light nuclei where the effects of electrostatic repulsion are smallest. Heisenberg assumed that only Coulomb forces acted between protons, and he argued that the neutron–neutron interaction must be small compared with the neutron–proton interaction in order to yield values of the ratio N/Z consistent with experimental data. Later it was found, however, that there was also a nuclear force acting between protons. According to Heisenberg's argument this would suggest an approximately equal force between neutrons and this led to the hypothesis of *charge symmetry* which means that the neutron–neutron and proton–proton forces are exactly equal except for Coulomb forces; or, equivalently, that all non-electrical forces are completely symmetrical between neutrons and protons. From this hypothesis follow simple quantitative predictions which can be tested experimentally. In

Section 4.1 of this chapter we look at its first application in 1936 to the prediction of binding energy differences between mirror nuclei.

At about the same time the first accurate experimental results on proton–proton scattering were published by Tuve, Heydenberg and Hafstad (1936) and analysed by Breit, Condon and Present (1936) (cf. Section 3.5). The analysis showed that the strengths of the neutron–proton and the proton–proton interactions were approximately the same if the ranges were taken to be equal. On the basis of this result Breit, Condon and Present proposed that the two interactions were exactly equal if Coulomb effects were neglected. This assumption taken together with charge symmetry gives the hypothesis of *charge independence*: Nuclear interactions are the same between all pairs of nuclear particles (except for Coulomb forces).

The wave function of a pair of protons or a pair of neutrons must be antisymmetric for exchange of space and spin coordinates because of the Pauli principle, whereas the wave function of a neutron and proton can be either symmetric or antisymmetric. The charge independence hypothesis implies that the force between a neutron and a proton in an antisymmetric state is the same as the force between two protons or two neutrons in the same state. Charge independence says nothing about the force between a neutron and a proton in a symmetric state, as a pair of protons or neutrons cannot exist in such a state.

The hypotheses of charge symmetry and charge independence are particularly interesting for the following three reasons. The first is historical. Most of the theoretical predictions had been derived before 1940, but they were not tested until ten years later. Serious experimental investigations began in about 1950 and the main predictions were verified during the next three or four years. The reason for this delay was probably that adequate experimental techniques did not exist in the late 1930's and were not developed until after the second world war. Secondly, these hypotheses provide the most direct link between nuclear physics and elementary particle physics. The conservation laws of isobaric spin and strangeness in elementary particle physics are direct extensions of the charge independence hypothesis in nuclear physics. The third point is that the hypotheses can be tested by studying properties of complex nuclei. This is almost the only field in which a study of complex nuclei can give quantitative information about the nucleon–nucleon interaction.

The reason for this last feature is that charge independence and charge symmetry imply a symmetry in the equations of motion of nucleons in a nucleus. It is a characteristic both of classical and quantum systems that physical symmetries lead to conservation laws. For example, if the laws of motion are invariant for translations of the origin of the coordinate system, then the total momentum is conserved. Similarly, invariance under rotations implies conservation of angular momentum and invariance with respect to changes of the origin of time implies conservation of energy. In quantum mechanics a symmetry introduces new quantum numbers which are eigenvalues of the conserved quantities. These eigenvalues can be used to classify states of the system irrespective of its complexity. Thus, all states of a nucleus have a definite total angular momentum, and angular momentum is conserved in reactions no matter how strong and complicated the forces are, provided only that they are invariant with respect to rotations. Charge independence and charge symmetry are new physical symmetries which lead to new conservation laws and associated quantum numbers. It is possible to use these conservation laws in order to make simple predictions about the properties of complex nuclei and to derive selection rules for nuclear reactions which may be tested experimentally.

The implications of charge independence for the properties of nuclear states were investigated by Wigner in 1937 and 1939. Selection rules were derived by Adair, and Kroll and Foldy in 1952. We include these last two papers in the reprint section of this book. Together they clarified the simple but important logical distinction between charge symmetry and charge independence.

If charge independence is true, also charge symmetry is true. Thus, any prediction which follows from the assumption of charge symmetry will also follow from charge independence, but certain predictions of charge independence do not follow from charge symmetry. In order to have a test of charge independence, it is necessary to find an effect that is not predicted from charge symmetry. For this reason it is harder to find definite tests of charge independence than of charge symmetry. There are fewer experiments that will do the job.

4.1 Binding Energies of Mirror Nuclei

Curie and Joliot discovered artificial radioactivity in 1934, and soon many radioactive light nuclei were produced and studied. Soon afterwards Fowler, Delsasso and Lauritsen (1936) investigated the positron decay of a number of elements of the type with Z protons and $Z-1$ neutrons into the "mirror nucleus" with $Z-1$ protons and Z neutrons. In their experiments the radioactive nuclei were produced by bombarding elements of low atomic weight with 1 MeV deuterons. They measured the maximum energy of positrons produced in the decay and used their results to determine the binding energy differences ΔW between pairs of mirror nuclei. In discussing their results they pointed out that if one assumed charge symmetry, then the change in binding energy from the parent to the daughter nucleus should be accounted for entirely by the electrostatic repulsion between the protons. If the density of nuclear matter inside the nucleus were approximately constant and if the nuclear volume were proportional to the atomic weight A, then the electrostatic energy difference could be calculated from classical electrostatics. They suggested therefore that the binding energy difference should be given by the formula

$$\begin{aligned} \Delta W &= (Z-1)e^2\,(r^{-1})_{av} \\ &= \frac{6e^2}{5r_0}(Z-1)A^{-\frac{1}{3}} \qquad (4.1) \\ &= 1{\cdot}20(Z-1)A^{-\frac{1}{3}}\ \text{MeV} \quad \text{if } r_0 = 1{\cdot}45\ \text{fm} \end{aligned}$$

where the nuclear radius $R = r_0 A^{\frac{1}{3}}$. The experimental results of Fowler, Delsasso and Lauritsen were consistent with this formula with a reasonable value of the nuclear radius. At about the same time Bethe applied the charge symmetry principle to determine the binding energy difference between H^3 and He^3. He showed that the experimental difference could be explained by the Coulomb force between the protons in He^3 (Bethe 1937, p. 146).

These experiments were continued and by 1941 a series of pairs of mirror nuclei with atomic weights up to $A = 41$ had been investigated and their binding energy differences measured. The results agreed well with the prediction based on the hypothesis of charge symmetry with $r_0 = 1{\cdot}45$ fm, almost the same as the value Gamow (1·4 fm) found from the α-decay of heavy nuclei. This state of affairs was disturbed in 1953 when other methods of measuring nuclear radii were developed. Fitch and Rainwater (1953) had measured the energies of x-rays from μ-mesic atoms and found that their results in atoms with $Z = 22$, 29, 51 and 82 were consistent with a nuclear radius $R = r_0 A^{\frac{1}{3}}$. Their value of $r_0 = 1{\cdot}2$ fm was, however, considerably less than the one needed to fit the binding energies of mirror nuclei. Electron scattering measurements have confirmed this smaller value for the radius of the nuclear charge distribution. When this discrepancy appeared it was recognized that the classical calculation of the electrostatic energy of nuclei was not correct. Cooper and Henley (1953) drew attention to some old calculations of Bethe (1937, 1938) on the contribution of exchange terms to the electrostatic energy and on the effect of the loose binding of the last proton. When the electrostatic energies of nuclei were calculated again including Bethe's corrections and with nuclear radii consistent with electron scattering experiments, the predicted differences in binding energy between pairs of mirror nuclei agreed with the experimental values.

If nuclear forces were symmetric between protons and neutrons, and if Coulomb forces were absent, the energies of all pairs of excited states of mirror nuclei as well as those of the ground states would correspond exactly. If the Coulomb repulsion between protons is included it shifts the relative energies of the ground states of mirror nuclei as discussed in the first part of this section. The Coulomb energy of a nuclear state depends mainly on its radius and is not very sensitive to details of its structure, so an excited state should be shifted by about the same amount as the ground state. Hence, energies of excited states relative to the ground state should be almost the same in pairs of mirror nuclei. Corresponding pairs should have the same spins and parities. This prediction has been verified in many cases. We show the low excited states of C^{11} and B^{11} in Fig. 7a as an example.

Although this prediction had been made in 1937 by Wigner there were no systematic experimental investigations until 1949. The experimental situation was reviewed in 1948 by Lauritsen, Fowler and Lauritsen, and the following picture emerges from their article. Despite the fact that a considerable variety of techniques were available for producing nuclear reactions to study excited states of nuclei, the application of any particular method was restricted by the energy attainable, the availability of a suitable target nucleus and the cross section for the reaction. These restrictions varied more or less at random from one nucleus to another and, as a result, many excited states were known in some nuclei and none at all in others.

In 1949, several experimental groups began to search for mirror states. The first positive evidence for a mirror state was found in Be^7. Brown, Chao, Fowler and Lauritsen† (1949) discovered an excited state at an energy of 434 keV by observing α-particles produced in the $B^{10}(p, \alpha)Be^7$ reaction. This state corresponded quite closely in energy to a well-known state in Li^7 at 479 keV. At about the same time disturbing evidence against the theory appeared. Two low excited states of C^{13} were known with excitation energies of 3·09 and 3·91 MeV. In 1949 Van Patter found resonances in the $C^{12}(p, \gamma)N^{13}$ reaction which indicated excited states of N^{13} at 2·29 and 3·48 MeV. He emphasized that energy levels of these mirror nuclei did not correspond, as would be expected if the proton–proton and neutron–neutron forces were equal.

Despite this discouraging case, evidence supporting charge symmetry continued to appear and in 1951 a whole sequence of excited states were measured in C^{11} and B^{11} with energies corresponding to within 200 keV. Nowadays, there is accurate data on the energies of the low excited states of many light nuclei. The worst deviation from the prediction of charge symmetry still occurs in N^{13} and C^{13} (cf.

† Fowler and Lauritsen had also collaborated in the first mirror binding energy experiment in 1936.

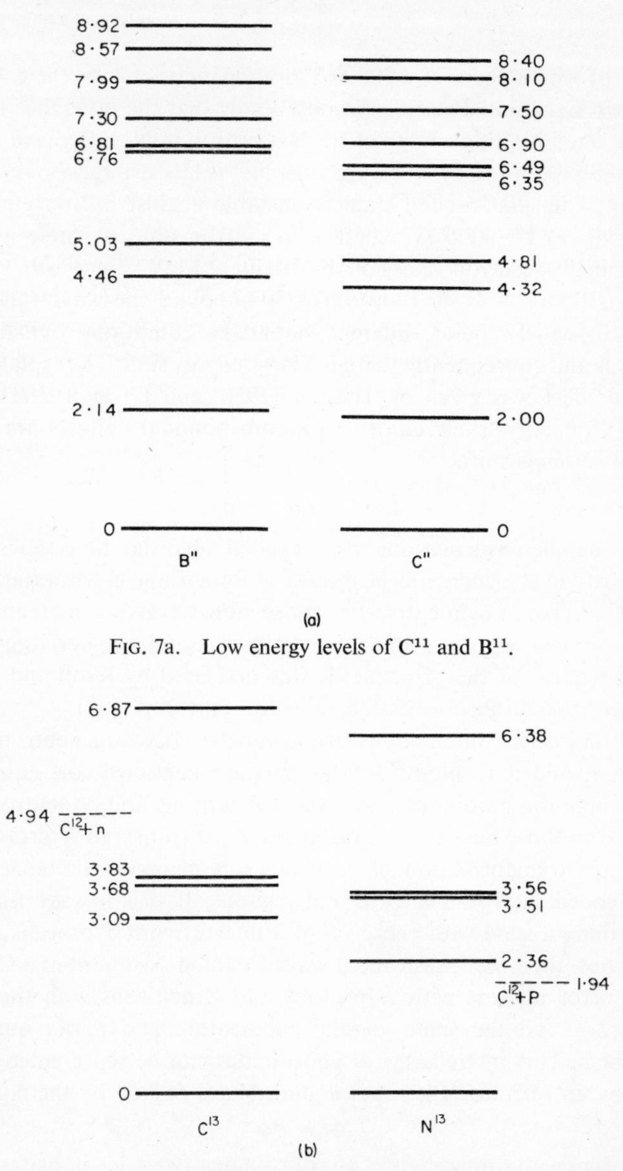

FIG. 7a. Low energy levels of C^{11} and B^{11}.

FIG. 7b. Low energy levels of C^{13} and N^{13} showing the thresholds C^{13} C^{12} + n, and N^{13} C^{12} + p.

Fig. 7b) where there is a 700 keV discrepancy between the energies of their first excited states. We now know that the difference is due to the small binding energy of the last proton in N^{13}. Because of the Coulomb repulsion between protons, N^{13} is less strongly bound than C^{13}, and its first excited state is unstable against proton emission (to $C^{12}+p$) by 400 keV whereas its mirror state is stable against breakup by neutron emission by about 2 MeV. Since one mirror state is bound and its counterpart is unbound the wave functions have to satisfy quite different boundary conditions outside the nucleus and consequently there is a large energy shift. The explanation of the effect was given by Thomas (1950) and Ehrman (1951) and shifts in energy levels due to Coulomb boundary effects are often called Thomas shifts.

4.2 Charge Parity

No detailed calculations were needed in order to establish the similarity of the energy level spectra of mirror nuclei from the hypothesis of charge symmetry. In this section we give a mathematical expression of charge symmetry which is useful for deriving other consequences of the principle. It was first used by Kroll and Foldy in their discussion of selection rules (cf. Part 2, p. 208).

If the nuclear interactions are symmetric between neutrons and protons, and if Coulomb interactions are neglected, the equations describing the motion of a system of Z protons and N neutrons are related to those describing a system of Z neutrons and N protons by a simple transformation of variables interchanging the space and spin coordinates of neutrons and protons. If ϕ is a wave function describing a state with energy E of a nucleus with Z protons and N neutrons, then the transformed wave function ϕ' describes a state of the mirror nucleus with N protons and Z neutrons with the same energy E and the same angular momentum and parity quantum numbers. This interchange of coordinates can be represented formally by an operator P so that ϕ and ϕ' are related by the equation

$$\phi' = P\phi$$

Performing the interchange of coordinates twice leads back to the original wave function; thus $P^2 = 1$.

A special situation occurs in a self-conjugate nucleus with $N = Z$ Then the state ϕ' is in the same nucleus as the state ϕ, and because it has the same energy it must be the same state. Thus

$$\phi' = P\phi = p\phi$$

where p is a numerical factor. As $P^2 = 1$ so $p^2 = 1$ or $p = \pm 1$. The states of such a nucleus divide into two classes: Those with $p = 1$ and those with $p = -1$. We shall see that states belonging to different classes have different physical properties.

Inclusion of Coulomb interactions shifts the relative energies of corresponding mirror states and also modifies the wave functions. These modifications are small, but not zero and while predictions following from charge symmetry remain almost true they are not exactly so and small deviations must be expected.

4.3 Charge Independence in Nucleon–Nucleon Scattering

The first proton–proton scattering experiments of Tuve, Heydenberg and Hafstad (1936) and the analysis of Breit, Condon and Present (1936) established the existence of a short range attractive interaction between protons comparable in strength with the neutron–proton interaction. On the basis of this evidence Breit, Condon and Present suggested that the proton–proton and neutron–proton interactions might be identical in $^1S_0$ state of relative motion, if Coulomb forces were neglected. This in turn lead to the hypothesis of charge independence that the same forces act between all pairs of nuclear particles. The suggestion was first made in 1936, and during the next fifteen years the experimental measurements of proton–proton and neutron–proton scattering cross sections were improved in accuracy so as to test its validity. The most accurate information we have now is summarized in the scattering length a_{pp} and effective range r_{0pp} for proton–proton scattering

$$a_{pp} = -7{\cdot}784 \pm 0{\cdot}030 \text{ fm} \qquad r_{0pp} = 2{\cdot}73 \pm 0{\cdot}08 \text{ fm}$$

and the corresponding quantities for singlet neutron–proton scattering

$$a_s = -23{\cdot}78 \pm 0{\cdot}035 \text{ fm} \qquad r_{0s} = 2{\cdot}670 \pm 0{\cdot}023 \text{ fm}$$

The two sets of parameters cannot be compared directly because of the Coulomb interaction between protons. Blatt and Jackson (1950) have shown that, assuming charge independence (i.e. if the same nuclear potential is used to calculate the singlet neutron–proton and the proton–proton scattering parameters) the two effective ranges turn out to be about the same, whereas the scattering lengths are related approximately by the formula

$$a_{pp}^{-1} = a_s^{-1} + \frac{1}{R}(\log(r_0/R) + 0{\cdot}33) \tag{4.2}$$

In this equation $R = \hbar/Me^2 = 2{\cdot}88$ fm is a characteristic Coulomb length for protons (i.e. it is the Bohr radius of a proton bound to a unit charge) and r_0 is the effective range of the interaction. The two effective ranges are indeed almost equal. If the measured value of a_{pp} is used to estimate the singlet scattering length a_s for neutron–proton scattering from equation (4.2), one obtains a value of about -17 fm instead of $-23{\cdot}8$ fm. There are two possible explanations for this discrepancy: It might be due to the inadequacy of equation (4.2) as this is only approximate. There is no way of comparing neutron–proton and proton–proton scattering which is completely independent of the shape of the interaction potential. If two potentials with different shapes and with strengths and ranges adjusted to fit proton–proton scattering are used to calculate the singlet neutron–proton scattering length, they will not produce exactly the same result, although in both cases the result will be somewhere near the prediction of equation (4.2). It is even possible to choose the shape of the potential so that the charge independence hypothesis is exactly true, i.e. to find a potential which reproduces the experimental values of both sets of low energy parameters (Preston and Shapiro 1956). On the other hand, it is more likely that the charge independence hypothesis is not exactly true and that the neutron–proton interaction is a little stronger than the proton–proton interaction. The calculated differences in strength vary from zero to about 3 per cent depending on the shape of the potential.

4.4 Isobaric Spin

Heisenberg introduced the isobaric spin as a means of describing a system made up of two kinds of particles. In his work the formalism implied no physical assumptions, and he could just as well have done without it. When charge independence was postulated Cassen and Condon (1936) realized that isobaric spin provided a natural way of describing the consequences of that hypothesis. They showed how the method could be applied to discuss states of the deuteron in a clear and simple paper which we reproduce in Part 2 of this book. They did not, however, work out more general consequences of the theory or show how it could be used for more complex nuclei. This was done by Wigner (1937) in a very important paper which also included the first work on a general method of classifying nuclear states called supermultiplet theory. This theory was much more involved than isobaric spin theory, and Wigner's paper was too difficult for most experimental and many theoretical physicists, and the simple consequences of charge independence were not widely recognized until many years later.

In Chap. III we saw that Heisenberg introduced the operators τ^{ξ}, τ^{η} and τ^{ζ} defined in the same way as the Pauli spin matrices (cf. Part 2, p. 146). In a system of A nucleons the operators

$$T_{\xi} = \tfrac{1}{2}\textstyle\sum_i \tau_i^{\xi}, \quad \text{etc.}$$

are analogous to the components of the total spin operator $\mathbf{S}$ of a system of electrons, and they obey the same commutation relations as the angular momentum operators. The mathematical properties of the isobaric spin operators can all be derived from these commutation relations and must be identical with those of the angular momentum operators. In particular, a state can be an eigenstate of both T_{ξ} and $\mathbf{T}^2 = T_{\xi}^2 + T_{\eta}^2 + T_{\zeta}^2$. The possible values of $\mathbf{T}^2$ are $T(T+1)$ where T is an integer or half integer. For a given value of T, T_{ζ} can take $2T+1$ values, namely $T, T-1 \ldots, -T$. We must interpret this formalism and show its relevance to the charge independence hypothesis.

The operator τ_i^{ζ} has the value 1 if particle i is a proton and -1 if it is a neutron. Hence the operator $\frac{1}{2}\sum_i (1+\tau_i^{\zeta}) = \frac{1}{2}A + T_{\zeta}$ is equal to the total number Z of protons in the nucleus. Thus

$$T_\zeta = Z - \tfrac{1}{2}A = \tfrac{1}{2}(Z - N)$$

This means that the eigenvalue of T_ζ determines the proton excess of a nuclear state. All states of the same nucleus must be eigenstates of T_ζ with the same eigenvalue. Also T_ζ must be an integer if A is even and a half-odd integer if A is odd. In the following discussion we use a more symmetric notation with $T_\zeta = T_0$.

The matrices τ^ξ and τ^η were introduced by Heisenberg in order to represent exchange forces mathematically. They both change a proton into a neutron and vice versa. More convenient matrices are the combinations $\tau^\pm = \tau^\xi \pm i\tau^\eta$. If the ith nucleon is a proton τ_i^+ changes it into a neutron; if it is a neutron already it transforms the wave function to zero. The operator τ_i^- has the opposite behaviour. Thus $T_+ = \frac{1}{2}\sum_i \tau_i^+$ transforms a wave function ϕ describing a nuclear state with Z protons and N neutrons into a wave function $\phi_+ = T_+\phi$ describing a state with $Z+1$ protons and $N-1$ neutrons. T_0 is increased by 1. Similarly, the state $\phi_- = T_-\phi$ describes a state with $Z-1$ protons and $N+1$ neutrons. In other words, the operators T_+ relate states in a given nucleus to states in neighbouring nuclei with the same atomic weight (isobaric nuclei).

Fermi (1934) used the operator $T_\pm$ in his theory of β-decay. In electron- or positron-decay a nucleus changes into a neighbouring isobaric nucleus, and the operators $T_\pm$ provide a convenient mathematical representation of this process.

The physical properties of the operators $T_\pm$ are reflected in the operator commutation rules

$$[T_0, T_\pm] = \pm T_\pm$$

If ϕ is an eigenstate of T_0 with eigenvalue t_0 and $\phi_+ = T_+\phi$ then

$$\begin{aligned} T_0\phi_+ = T_0 T_+ \phi &= [T_0, T_+]\phi + T_+ T_0 \phi \\ &= T_+\phi + t_0 T_+ \phi \\ &= (t_0 + 1)\phi_+ \end{aligned}$$

i.e. ϕ_+ is an eigenstate of T_0 with eigenvalue t_0+1. Or, T_+ has changed a neutron into a proton. With some states it may not be possible to change a neutron into a proton without violating the

Pauli exclusion principle. In that case $T_+\phi = 0$. The operators $T_\pm$ commute with $\boldsymbol{T}^2$. If ϕ is an eigenstate of $\boldsymbol{T}^2$ with eigenvalue $T(T+1)$, so is $T_\pm\phi$. If $T_0\phi = T\phi$ then $T_+\phi = 0$ because T is the maximum value of T_0.

Physically the hypothesis of charge independence means that nuclear forces between all pairs of nucleons are equal. In the mathematical formalism this is equivalent to the statement (Wigner 1937) that the Hamiltonian describing the nucleus does not depend on the isobaric spin coordinates or more precisely, that the Hamiltonian operator can be written in terms of space and spin variables alone and does not differentiate between neutrons and protons. This does not mean that it cannot be written in another way, and there are often many equivalent ways of writing the same operator.

In Chap. II we discussed the forces introduced by Heisenberg, Majorana and Wigner. The most general central interaction $V(r)$ is a linear combination of an ordinary potential $V_W(r)$ of the type introduced by Wigner, a Heisenberg and a Majorana exchange interaction $V_H(r)P_M P_\sigma$ and $V_M(r)P_M$ and a fourth spin exchange interaction first used by Bartlett (1936), $V_B(r)P_\sigma$,

$$V = V_W(r) + V_B(r)P_\sigma + V_M P_M + V_H(r)P_M P_\sigma$$

The operators P_M and P_σ are the coordinate and spin exchange operators. If the exchange operators P_M and P_σ exchange the coordinates and spins of both like and unlike particles† the interaction V is charge independent because it does not differentiate between neutrons and protons, neither does it depend on the isobaric spin coordinates.

We proved in Chap. II that $P_M P_\sigma = -P_c$, where P_c is the charge exchange operator. P_c can be expressed in terms of the isobaric spin coordinates $\boldsymbol{\tau}_1$ and $\boldsymbol{\tau}_2$ of the interacting nucleons

$$P_c = \tfrac{1}{2}(1 + \boldsymbol{\tau}_1 . \boldsymbol{\tau}_2)$$

(cf. Cassen and Condon, Part 2, p. 193). Similarly

$$P_\sigma = \tfrac{1}{2}(1 + \boldsymbol{\sigma}_1 . \boldsymbol{\sigma}_2)$$

† The exchange operators P'_M and P' used in Chap. II exchanged the coordinates and spins of unlike nucleons only and were equal to zero for a pair of like nucleons.

and

$$P_M = -\tfrac{1}{4}(1+\boldsymbol{\sigma}_1 . \boldsymbol{\sigma}_2)(1+\boldsymbol{\tau}_1 . \boldsymbol{\tau}_2)$$

Hence the same interaction V may be written in another way

$$V = V_1(r) + V_2(r)\boldsymbol{\sigma}_1 . \boldsymbol{\sigma}_2 + V_3(r)\boldsymbol{\tau}_1 . \boldsymbol{\tau}_2 + V_4(r)(\boldsymbol{\sigma}_1 . \boldsymbol{\sigma}_2)(\boldsymbol{\tau}_1 . \boldsymbol{\tau}_2)$$

where the V_1, V_2, V_3, V_4 are linear combinations of V_W, V_M, V_B and V_H. This interaction depends explicitly on the isobaric spin coordinates, but it is completely equivalent to the first form which depends only on space and spin coordinates because of the antisymmetry of the total nuclear wave function for exchange of space, spin and isobaric spin coordinates of two nucleons. The second form containing isobaric spin coordinates explicitly appears in potentials derived from the meson theory of nuclear forces.

The above definition of charge independence means that the operators T_0, $T_\pm$ and $\mathbf{T}^2$ all commute with a charge independent Hamiltonian operator. In consequence the nuclear states, which are eigenstates of the Hamiltonian, are also eigenstates of $\mathbf{T}^2$ with some eigenvalue $T(T+1)$. Thus each state of a nucleus has a quantum number T in addition to its energy, the angular momentum and parity quantum numbers. The new quantum number is called the isobaric spin. It can have integral values when A is even and half-odd integral ones when A is odd. Supposing a nucleus with Z protons and N neutrons has a state energy E, spin J, parity π and isobaric spin T. Because the operator T_+ commutes with the Hamiltonian, the angular momentum and parity operators and $\mathbf{T}^2$, the state $\phi_+ = T_+\phi$ is also a state with energy E and with the same angular momentum, parity and isobaric spin as ϕ. Hence, the state ϕ may have counterparts with the same quantum numbers in neighbouring isobaric nuclei.

4.5 Isobaric Multiplets

If nuclear forces are charge independent the theory of isobaric spin predicts the existence of isobaric multiplets, i.e. groups of states with the same quantum numbers in isobaric nuclei. These were first discussed by Wigner in 1937.

We saw in the last section that charge independence implies that nuclear states have an isobaric spin quantum number T in addition to the angular momentum and parity. If a state of some nucleus has an isobaric spin number T then there are altogether $2T+1$ states with the same energy, but with values of T_0 ranging from T to $-T$. Because $T_0 = \frac{1}{2}(Z-N)$ these are states in $2T+1$ different nuclei with the same A but with $N-Z$ ranging from $2T$ to $-2T$. Several isobaric nuclei should have energy levels with the same binding energy. In particular, if a nucleus with a certain value of $|N-Z|$ has an energy level with a specific binding energy each isobaric nucleus

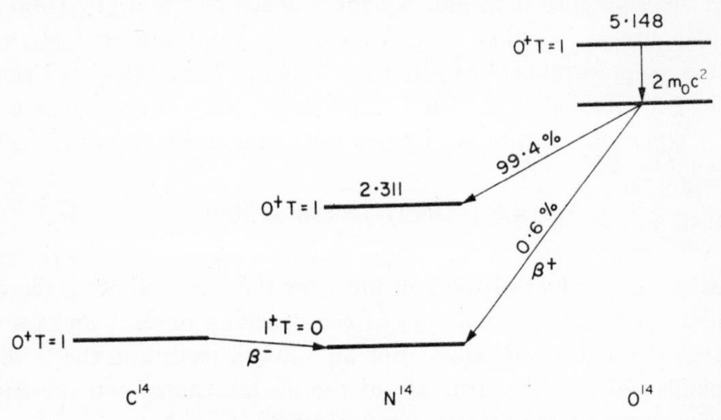

FIG. 8. The β-decay of C^{14} and O^{16}.

with an equal or smaller value of $|Z-N|$ should have a level with the same binding energy. A level with $T=0$ appears only in a self-conjugate nucleus with $N=Z=A/2$. A level with $T=1$ can occur in a self-conjugate nucleus and also in the neighbouring nuclei with $Z=A/2\pm 1$ and $N=A/2\mp 1$ and so on. All levels of the same multiplet have the same energy, angular momentum and parity.

The nuclear forces are not exactly charge independent if only because of the Coulomb repulsion between protons. The Coulomb forces shift the binding energy of nuclear states as was discussed in Section 4.1, but the Coulomb energy can be subtracted out using formula (4.1), and the energies of corresponding states in isobaric nuclei should be almost the same when this has been done.

The first successful comparison was made in the isobaric triplet C^{14}, N^{14}, O^{14} by Sherr, Muether and White (1948). They studied the positron decay of O^{14} and found that it led to an excited state of N^{14} which decayed emitting 2·3 MeV γ-rays. The positron end point was measured to be 1·8 MeV. From this experiment they concluded that N^{14} had an excited state at 2·3 MeV and that the difference in atomic weight between N^{14} and O^{14} was 5·1 MeV. They suspected that the ground states of C^{14} and O^{14} and the 2·3 MeV state of N^{14} were three states of an isobaric triplet ($T = 1$), and to test their conjecture they calculated the position of the excited state in N^{14} and the difference in atomic weight between N^{14} and O^{14} from the known atomic weights of C^{14} and N^{14}. Coulomb energies were found using formula (4.1) with $r_0 = 1{\cdot}44$ fm. Their calculated values were 2·39 MeV and 5·12 MeV respectively, both in good agreement with experimental values. Figure 8 shows the low states of C^{14}, N^{14} and O^{14}.

4.6 β-Decay Transition Rates

A β-decay transition rate is conveniently measured by its ft value. This is the product of the half-life t for the decay, times a factor f which depends on the charge of the decaying nucleus and on the energy emitted in the decay. For an allowed transition the ft value depends only on the structure of the nuclear states and the decay mechanism and not on the energy released,

$$ft = \frac{2\pi^3(\hbar/mc^2)\log_e 2}{[g_V^2|M_F|^2 + g_A^2|M_\sigma|^2]}$$

In this equation M_F and M_σ are the Fermi† and Gamow–Teller matrix elements and g_V and g_A are the corresponding coupling constants ($g_V = 2{\cdot}95 \times 10^{-12}$ and $g_A/g_V = 1{\cdot}24$). Measurements show that the ft values are much smaller than the average (i.e. the transition matrix elements M_F and M_σ are large) for transitions between corresponding states of mirror nuclei, and for this reason such transitions are called *super-allowed*. The large transition rates are explained by the similarity in structure of the initial and final states. Nordheim

† $M_F = (f|\, T_\pm \,|i)$, $M_\sigma = (f|\, \Sigma_j \sigma_j \tau_j^\pm \,|i)$ where $|i)$ and $|f)$ are the initial and final states.

and Yost pointed this out in 1937 and suggested the absolute magnitude of the lifetimes of certain emittors could be predicted on the basis of Fermi's theory. Wigner took the problem up and showed, provided nuclear forces are charge independent, that the Fermi matrix element M_F for a β-transition between states of the same T-multiplet could be predicted from the Fermi coupling constant and the isobaric spin quantum numbers of the initial and final states without any further knowledge of the nuclear wave function. The Fermi matrix element should be zero for transitions between states belonging to different charge multiplets. These predictions are difficult to test because in most cases the Gamow–Teller type interaction also contributes to a β-decay transition matrix element, and its contribution cannot be calculated without more detailed information about the nuclear wave functions. There are, however, just a few transitions where the Gamow–Teller transition is forbidden ($M_\sigma = 0$) and the only contribution comes from the Fermi interaction. This happens when both the initial and final state have zero spin and the same parity. The best known example (Fig. 8) is the transition between the 0^+ ground state of O^{14} to the 0^+ state in N^{14} with an excitation energy of 2·3 MeV above the N^{14} ground state. These two states belong to the same isobaric multiplet and Wigner's formula for the transition matrix element is

$$\begin{aligned} M_F &= (T, T_0^f | T_\pm | T, T_0^i) \\ &= [T(T+1) - T_0^i T_0^f]^{\frac{1}{2}} \end{aligned}$$

where T is the isobaric spin of the multiplet and T_0^i and T_0^f are the components of the initial and final state. For the O^{14}, N^{14} transition $T = 1$ and $T_0^f = 0$ so that $M_F = \sqrt{2}$. The formula for the Fermi matrix element was given by Wigner in 1939, but efforts to test it experimentally were made only after 1953.

In Table 3 we give recent *ft* values for the most carefully measured transitions between states with zero spin and positive parity belonging to $T = 1$ isobaric multiplets. The *ft* values should all be the same, if the Fermi theory is correct and if nuclear forces are charge independent. The experimental values differ from each other by less than 1 per cent.

TABLE 3

ft-Values of Pure Fermi Transitions

	O^{14}	Al^{26}	Cl^{34}	V^{46}	Co^{54}
(*a*)	3066 ± 10	3015 ± 12	3055 ± 20	3011 ± 25	2966 ± 18
(*b*)	3173	3135	3176	3172	3165

(*a*) Experimental values quoted from Freeman *et al.* (1964).
(*b*) With corrections for nuclear size, electron screening and radiative corrections.

There are some examples of β-transitions between states with zero spin and the same parity where the initial and final states belong to different charge multiplets. If the charge independence hypothesis were exactly true these transitions should be forbidden. Coulomb effects and a possible small charge dependence of the nuclear force (cf. Section 4.2) would relax this selection rule somewhat, and the transitions would occur slowly (i.e. with large *ft* values). There are two examples of this kind of transition in the decay sequence ${}_{32}Ge^{66} \rightarrow {}_{31}Cu^{66} \rightarrow {}_{30}Zn^{66}$. The *ft* values of these transitions are a factor of about 10^4 larger than the ones given in Table 3 and the prediction of charge independence is well verified (Alford and French 1961).

To end this section we give a case where simple predictions following from charge symmetry are not verified. The most striking discrepancy occurs in the decays

$$C^{14} \rightarrow N^{14} + e^- + \tilde{\nu}$$

$$O^{14} \rightarrow N^{14} + e^+ + \nu$$

leading to the *ground* state of N^{14} (Fig. 8). These are mirror transitions and should have equal *ft* values. The experimental *ft* values for the decay of C^{14} and O^{14} however, are $1 \cdot 12 \times 10^9$ and $2 \cdot 0 \times 10^7$ respectively. They differ by a factor of 50.

At first sight there is a serious discrepancy between the prediction of charge symmetry and the experimental result, but a closer investigation shows that these particular transitions are very sensitive to small changes in the wave functions. The deviation from charge symmetry due to Coulomb forces between protons is probably sufficient to explain the difference in the observed ft values (Ferrell and Visscher 1957). The explanation is as follows.

The β-transitions from the C^{14} and O^{14} ground states (spin and parity $J^{\pi} = 0^{+}$) to the N^{14} ground state ($J^{\pi} = 1^{+}$) are allowed by Gamow–Teller selection rules, but their ft values are abnormally large. A simple estimate based on the shell model would predict an ft value between 10^3 and 10^4 while the measured values are $1{\cdot}12 \times 10^9$ and $2{\cdot}0 \times 10^7$ respectively. (The half life of C^{14} should be about one day instead of 1000 years.) In other words the β-decay matrix elements (inversely proportional to the square root of the ft value) are about two orders of magnitude smaller than expected. The small values seem to be due to an accidental cancellation in the β-decay matrix element. This fact makes the relative transition rates of the O^{14} and C^{14} sensitive to small deviations from charge symmetry. In the following paragraph we shall estimate the difference in the ground state wave functions of O^{14} and C^{14} needed to produce the observed difference in β-decay matrix elements.

Let ϕ_c and ϕ_0 be the ground state wave functions of C^{14} and O^{14} respectively. If the nuclear forces were exactly charge symmetric then $\phi_0 = P\phi_c$, i.e. the wave function of O^{14} would be obtained from that of C^{14} by interchanging neutron and proton coordinates. When Coulomb forces are taken into account the wave function ϕ_0 is no longer exactly the "mirror" image of ϕ_c but contains some small admixture of other states. In general the wave functions ϕ_0 and ϕ_c are related by

$$\phi_0 = (P\phi_c + \alpha\chi)/(1+\alpha^2)^{\frac{1}{2}}$$

where the admixed wave function χ is orthogonal to $P\phi_c$ and the amplitude α measures the amount of admixture. If α is small, the normalization constant $(1+\alpha^2)^{\frac{1}{2}} \simeq 1$. There is a corresponding relation between the β-decay matrix elements M_0 and M_C of O^{14} and C^{14}

$$M_0 = M_C + \alpha M_\chi \tag{4.3}$$

where M_χ is the α-decay matrix element of the admixed state χ of O^{14} to the ground state of N^{14}. (If ϕ_N is the ground state wave of N^{14} and B is the β-decay operator then $M_0 = (\phi_0|B|\phi_N)$ and $M_\chi = (\chi|B|\phi_N)$. The magnitudes of M_0 and M_C obtained from experimental ft values are

$$M_C = 8{\cdot}3 \times 10^{-4}, \qquad M_0 = 6{\cdot}3 \times 10^{-3}$$

Therefore from equation (4.3)

$$|\alpha M_\chi| < 7 \times 10^{-3}$$

The admixed state would be expected to have a "normal" ft value of about $10^4 (|M_\chi| \doteqdot 0{\cdot}25)$, hence $\alpha \doteqdot 0{\cdot}03$.

In this abnormal example a deviation from charge symmetry in the wave functions of O^{14} and C^{14} corresponding to a 3 per cent admixture in amplitude would produce a difference of a factor of 50 in the β-decay transition rates. Is such an admixture consistent with the hypothesis of charge symmetry? In order to answer this question it is necessary to calculate the effect of the Coulomb forces on the wave function by perturbation theory or some other approximate method. The calculations require a fairly detailed knowledge of the nuclear wave functions and each case must be examined individually. Ferrell and Visscher made such an analysis for C^{14} and O^{14} and concluded that electromagnetic effects could explain the deviation from charge symmetry in the β-decay transition rates.

Although the C^{14} and O^{14} β-decays provide the most striking example, deviations from the prediction of charge symmetry seem to be the general rule for "mirror" β-transitions, e.g. $ft(N^{12})/ft(B^{12}) = 1{\cdot}11 \pm 0{\cdot}01$.

4.7 Nuclear Reactions

By 1951 data was beginning to accumulate on the energies of states in mirror nuclei, and the predictions of charge symmetry seemed to be confirmed. Charge independence had predicted the existence of isobaric multiplets, but only two cases had been found. A way of identifying the members of an isobaric multiplet was needed and became available when it was realized that the postulates of charge symmetry and charge independence lead to selection rules in nuclear reactions and for electromagnetic transitions.

The first paper on this subject was published in 1952 by Adair (cf. Part 2, p. 202). Charge independence implies that the total isobaric spin of a system must be conserved in any process. In particular, in a nuclear reaction the total isobaric spin in the final state must be equal to its value in the initial state. The T_0 component of isobaric spin is also conserved, but this is equivalent to conservation of charge and leads to no new results. The selection rules are most simply explained by examples, and we shall look at three cases the first of which is also discussed by Adair. Others are considered in his paper (cf. Part 2, p. 208).

If two nuclei A and B have isobaric spins T_A and T_B, the possible values of the total isobaric spin T of the two nuclei A and B is restricted by the relation

$$|T_A + T_B| \geqslant T \geqslant |T_A - T_B| \tag{4.4}$$

We need this result for the following discussion:

(a) The reaction O^{16}(d, α) N^{14} leading to various excited states of N^{14}. The initial state consists of O^{16} +d. Both these nuclei have isobaric spin equal to zero, so relation (4.4) implies that the total isobaric spin T_i of the system before the reaction is $T_i = 0$. The final state consists of $\alpha + N^{14}$. The α-particle has zero isobaric spin, so the total iso-spin T_f of the final state is equal to T_N, the iso-spin of the particular state of the nitrogen nucleus which is being investigated. Because of conservation of isobaric spin $T_i = T_f$. Hence, $T_N = 0$ and the reaction can proceed only to states of N^{14} with isobaric spin zero. Groups of α-particles are observed leading to the ground state of N^{14} and to several excited states but not to the state at 2·3 MeV, confirming that this is at $T = 1$ state.

(b) The inelastic scattering of α-particles by N^{14}. Because the ground state of N^{14} and the α-particle both have zero isobaric spin a similar argument to the above example shows that the inelastic scattering can lead only to excited states with $T = 0$, and a transition to the state at 2·3 MeV ($T = 1$) should be forbidden.

(c) Inelastic scattering of protons by N^{14}. In the initial state the N^{14} nucleus and the proton have iso-spin zero and $\frac{1}{2}$ respectively. It follows from equation (1) that the total iso-spin of the initial state is

$T_i = \frac{1}{2}$. Hence, the total iso-spin of the final state must be $T_f = T_i = \frac{1}{2}$. The proton has iso-spin $\frac{1}{2}$ and both $T_N = 0$ and $T_N = 1$ are compatible with the inequality (4.4). There is no selection rule forbidding transitions to the 2·3 MeV $T = 1$ state of N^{14}. The transition to the 2·3 MeV state of N^{14} shows up strongly in the (p, p′) scattering but not in the (α, α') process.

We have remarked before that certain consequences of charge independence also follow from the weaker hypothesis of charge symmetry. This was not realized clearly at the time Adair wrote his paper, but was stressed shortly afterwards by Kroll and Foldy (cf. Part 2, p. 208). They showed that many of Adair's selection rules could be derived from charge symmetry alone. These two papers helped to clarify the distinction between the two hypotheses.

Charge symmetry leads to selection rules in a reaction

$$A + B \rightarrow C + D$$

if each nucleus is self-conjugate. Then each nuclear wave function has a definite charge parity p_A, p_B, p_C and p_D respectively. Conservation of charge parity implies that

$$p_A p_B = p_C p_D$$

(Charge parity is a multiplicative quantum number like space parity.) In the examples discussed above the ground states of O^{16}, d, α and N^{14} all have even charge parity ($p = +1$). In example (a) O^{16}(d, α) N^{14} the conservation law (a) of charge parity implies that $p_N = +1$, i.e. only excited states with even charge parity can be reached in the reaction.

The transition to the 2·3 MeV state is not observed which indicates that it has negative charge parity. A similar discussion holds for the second example, but charge parity says nothing about the third example because the proton is not self-conjugate. The selection rules in examples (a) and (b) follow from the hypothesis of charge symmetry. Hence, their verification by experiment does not provide a definite proof of charge independence. They are useful, however, for determining the isobaric spin of a state, because of the relation between charge parity and isobaric spin $p = (-1)^T$. Thus a measurement of the charge parity can determine whether the iso-spin is even or odd. This is often sufficient information to fix its value.

4.8 γ-Ray Transitions

Selection rules for γ-ray transitions were first investigated by Trainor (1952) and Radicati (1952). In this section we discuss the particular case of E1 (electric dipole) transitions where the selection rules are most stringent. The electric dipole operator is just

$$\mathbf{D} = \sum e_i \mathbf{r}_i$$

where $\boldsymbol{e}_i$ is the charge of the ith nucleon and $\boldsymbol{r}_i$ is its position. Following Heisenberg we can put $\boldsymbol{e}_i = \frac{1}{2}e(1+\tau_i^\zeta)$ or

$$\mathbf{D} = \tfrac{1}{2}e \sum \mathbf{r}_i + \tfrac{1}{2}e \sum_i \tau_i^\zeta \mathbf{r}_i \tag{4.5}$$

In the first term of equation (4.5) $\sum \boldsymbol{r}_i$ is proportional to the coordinates of the centre of mass of the nucleus. This part of the dipole operator can only change the motion of the nucleus as a whole and cannot cause excitation. Hence, the effective dipole operator is

$$\mathbf{D}' = \tfrac{1}{2}e \sum \tau_i^\zeta \mathbf{r}_i$$

If we interchange proton and neutron coordinates then $\tau_i^\zeta \rightarrow -\tau_i^\zeta$ and $\mathbf{D}' \rightarrow -\mathbf{D}'$, or the operator $\mathbf{D}'$ has negative charge parity. Thus, if the principle of charge symmetry holds an E1 transition in a self-conjugate nucleus can only occur between states with different charge parities.

Charge independence implies a selection rule $\Delta T = 0$ or 1 for all transitions. This selection rule is not very restrictive because isobaric spins of low excited states of light nuclei seldom differ by more than one. In a self-conjugate nucleus, however, the isobaric spin is related to the charge parity by the equation $p = (-1)^T$, hence a $\Delta T = 0$ transition is forbidden for E1 radiation because of the charge parity selection rule. There is thus a selection rule $\Delta T = 1$ for E1 radiation in a self-conjugate nucleus. To illustrate the effect of this selection rule we give some transition rates for E1 transitions in O^{16} in Table 4. The transition rates are given as the ratio of the observed γ-ray width of the energy level to the Weisskopf "single particle estimate"

TABLE 4

Some E1 γ-Transition Rates in O^{16}

Initial state energy (MeV)	Final state energy (MeV)	ΔT	Γ/Γ_W
7·12	0	0	4×10^{-4}
7·12	6·06	0	5×10^{-7}
9·58	0	0	5×10^{-6}
13·1	0	1	0·14
17·7	0	1	0·06
22·5	0	1	0·39

of the width. This takes out the energy dependence of the γ-ray transition probability. All transitions are from a spin one, odd parity initial state to a spin zero, even parity final state. Some are allowed by the selection rule $\Delta T = 1$ and some are forbidden. The latter all have very small matrix elements compared with the former. Many experiments testing the $\Delta T = 1$ selection rule in self-conjugate nuclei have been made by Wilkinson and we refer to a review article by him (Wilkinson 1958) which summarizes the experimental information on isobaric spin selection rules for γ-radiation.

We have seen in this chapter that the predictions of charge symmetry and charge independence are verified well by experiments but never absolutely so. There are always small deviations. Energy levels in mirror nuclei do not correspond exactly and transitions which are forbidden by isobaric spin selection rules occur slowly. In most cases these small departures from the predictions of the theory can be ascribed to the effects of Coulomb forces, but there may be small differences between the proton–proton, neutron–neutron and neutron–proton nuclear forces. The results of experiments of the type described in this chapter limit these differences in interaction strengths to less than about 3 per cent.

V

Nucleon–Nucleon Scattering at High Energies

IN Chap. IV we dealt with studies of light nuclei that provided much evidence in support of the charge independence and charge symmetry hypotheses. Other nuclear structure studies have given only qualitative information about nuclear forces. The low energy scattering experiments discussed in Chap. III could determine the strength and range of the nuclear interaction, and the deuteron quadrupole moment established the existence of tensor forces, but low energy experiments could give no more information than this. Most of our detailed knowledge of nuclear forces nowadays comes from high energy proton–proton and neutron–proton scattering experiments. There is very little direct information about the neutron–neutron force, but charge symmetry tells us that it is almost the same as the proton-proton force.

Two monographs on nucleon–nucleon scattering have been published recently. One by Wilson (1963) covers the experimental and phenomenological aspects and gives a complete summary of all the experimental data available at the time of publication. The other by Moravcsik (1963) introduces the theoretical problems and shows how the experimental results have been evaluated in terms of theoretical models. In this chapter we give a brief summary of the important work which is discussed in detail in these two books.

Nucleon–nucleon scattering experiments have been carried out with proton energies up to 26 GeV. We shall consider only experiments below or just above the threshold for production of π-mesons (290 MeV). Angular distributions of scattered particles have been measured at many energies between zero and about 300 MeV. Also polarization measurements have been made. Most of these have been simple polarization experiments in which the incident and target nucleons were unpolarized and the polarization of one of the scattered nucleons was measured by a second scattering. In the so-called triple scattering experiments both the incident and scattered nucleons were polarized.

5.1 Phase Shift Analysis

If a complete theory of nuclear forces was available, theoretical angular distributions and polarizations could be calculated from this theory and compared directly with experimental data. However, no such theory exists at present and in the absence of it, it is most convenient to summarize the experimental data by a set of phase shifts at each energy.

We considered the scattering of spinless particles by a central force in Chap. III and showed how the differential scattering cross section at a particular energy was determined by a set of phase shifts. Neutrons and protons have non-zero spins, but the phase shift method may be generalized and any experimental quantity can be expressed in terms of parameters analogous to phase shifts.

The phase shift method for spinless particles is based on the decomposition of the scattered wave into "partial waves" each with definite orbital angular momentum l. The scattering at a particular energy in the lth partial wave is determined completely by a phase shift δ_l.

It is still possible to make a decomposition into states with definite angular momenta even if the particles involved have spins. We shall discuss proton–proton scattering first. The Pauli exclusion principle allows only states that are completely antisymmetric in space and spin coordinates. Singlet states (with spin zero) must have even and triplet states must have odd orbital angular momentum. The possible angular momentum states in proton–proton scattering are

$$^1S_0,\ ^3P_{0,1,2},\ ^1D_2,\ ^3F_{2,3,4},\ ^1G_4,\ ^3H_{4,5,6}\cdots$$

where the subscripts indicate the total angular momentum J. In triplet states $S = 1$, and there are three possible J values for each L ($J = L+1$, L or $L-1$) while in singlet states $S=0$ and $J=L$. If we assume that total angular momentum and parity are conserved, a state with a certain spin and parity ($J\pi$) before the scattering has the same values for these quantum numbers after the scattering. (The parity is determined by the orbital angular momentum, $\pi=(-1)^L$.) In most cases this implies the final state must be the same as the initial one, and the scattering is determined by a phase shift $\delta_{J\pi}$. For example, there is only one state with orbital angular momentum $J=1$ and negative parity: This is $^3P_1$ and the scattering in this state is determined by a phase shift $\delta(^3P_1)=\delta_{1-}$. There are some cases, however, where there are two states with the same angular momentum and parity, e.g. $^3P_2$ and $^3F_2$. If the state of the incident particles is $^3P_2$, then the scattered state may be a mixture of the two states $^3P_2$ and $^3F_2$. The orbital angular momentum L is changed, but it does not have to be conserved if there are non-central forces such as tensor forces acting between the nucleons. In this situation the scattering is determined by a 2×2 unitary matrix. If the laws of nature are invariant with respect to changes in sign of the time coordinate this matrix is specified by three independent real parameters which are usually taken to be two phase shifts and one "mixing" parameter. The two most common ways of defining these phase shifts are discussed by Moravcsik (1963). If partial waves up to $L=3$, i.e. F waves, contribute to the scattering at a certain energy, there are nine parameters to be determined from the experiments (eight phase shifts and one mixing parameter). The differential cross section can provide at most four relations between these parameters, and polarization measurements are necessary to determine them all.

There is no restriction from the Pauli exclusion principle in neutron–proton scattering. States which are completely symmetric in space and spin coordinates are also allowed. These are

$$^3S_1, {}^1P_1, {}^3D_{1,2,3}, {}^1F_3 \ldots$$

If the charge independence hypothesis is true, the $^3P_1$ and $^1P_1$ states do not mix. Although they have the same angular momentum and parity they have different isobaric spins ($T=1$ and $T=0$

respectively). The $^3S_1$ and $^3D_1$ states can mix as we have already seen when discussing the deuteron quadrupole moment.

There are some difficulties in obtaining phase shifts from experimental data. The algebraic equations relating the phase shifts to experimental quantities are non-linear. Even if the experiments are very accurate there are usually several sets of phase shifts which fit the data equally well. With low energy scattering in Chap. III we saw, for example, that measurements of the cross section could determine the magnitude of the scattering length but not its sign. Experimental quantities are always products of two amplitudes. For this reason, changing the sign of all phase shifts will not alter an experimental quantity, and conversely, scattering experiments can never determine the overall sign of the phase shifts. This is the simplest ambiguity. There are several ways of selecting the correct solution from the several possible ones.

The following method can be used, if scattering experiments are performed over a range of energies from zero upwards. A phase shift analysis at any one energy will give more than one solution. These different solutions will vary continuously as the scattering energies change. We know the asymptotic behaviour of the phase shifts as the scattering energy goes to zero. Usually only one solution will vary with energy in an acceptable way and approach the correct limits for low energies.

If something is known theoretically about the interaction it may help to decide between the different solutions. We do not have a complete theory of nuclear forces, but we do know the form of the interaction between two nucleons from meson theory, if their separation is not too small (cf. Chap. VI). Classically, if a scattered particle has a large angular momentum, it passes a long way from the scatterer. The impact parameter $b = l\hbar/p$ gives an estimate of the distance of closest approach of the particles involved in the scattering (p is the momentum in the centre of mass system and $\hbar l$ is the angular momentum). If the orbital angular momentum is large enough the particles should only be influenced by the long range part of the interaction between them as they scatter off each other. It should be possible to predict the phase shifts of partial waves with large angular momentum from the theoretical forces for large separations. This method was first used in the phase shift analysis of nucleon–nucleon scattering in 1958.

The experimental data for proton–proton scattering have been analysed by Breit (1960) and Stapp (1963) who obtained statistically good phase shift representations over the entire energy range from

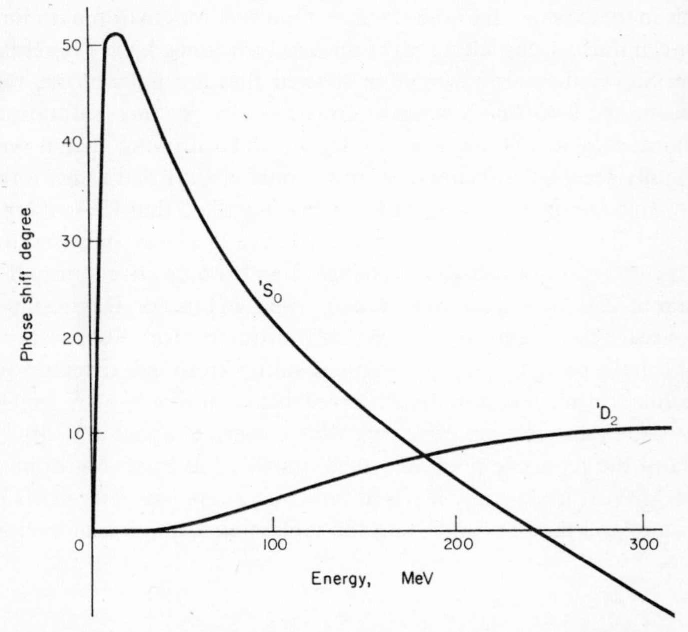

FIG. 9. Sketch of the $^1S_0$ and $^1D_2$ proton–proton phase shifts.

zero to 400 MeV. A similar analysis has been carried out for neutron–proton scattering assuming charge independence so that the $T=1$ phase shifts can be taken from the proton–proton scattering analysis (Breit 1960, 1962). The experimental data are not so accurate as for proton–proton scattering, hence the phase shifts are less reliable. To illustrate the results of these analyses we give a sketch of the $^1S_0$ and $^1D_2$ proton–proton phase shifts in Fig. 9.

5.2 The Repulsive Core

One very important qualitative fact which emerged from the results of nucleon–nucleon scattering experiments is the existence of a repulsive core in the nucleon–nucleon interaction.

When Heisenberg first discussed the saturation of nuclear forces and the constant density of nuclear matter, he concluded that nuclear forces must become strongly repulsive for small separations between nuclear particles, i.e. the nuclear force must have a repulsive core. Soon afterwards Majorana showed that a repulsive core was unnecessary and that exchange forces could produce saturation. Majorana's explanation was so elegant and satisfying that it was generally accepted as being true. It was only in 1951 that some of the early high energy scattering experiments suggested that Heisenberg's original idea might be correct (Jastrow 1951). We know now that the $^1S_0$ proton–proton phase shift changes sign from positive to negative at about 250 MeV laboratory kinetic energy (Fig. 9). This result is inconsistent with a nuclear force that is attractive for all separations between the particles. A negative phase shift corresponds to repulsion. In order to account for the observed behaviour of the $^1S_0$ phase shift there must be a repulsive core with a radius of about $0{\cdot}5 \times 10^{-13}$ cm and the repulsive potential energy must be at least several hundred MeV. It is generally believed now that a repulsive core exists in all states and that it is important for explaining saturation of nuclear forces.

VI
The Meson Theory of Nuclear Forces

As a result of the developments made in the quantum theory of radiation during the 1920's it was recognized that photons can be created or destroyed when they interact with matter. When an atom in an excited state makes a transition to a lower state emitting radiation, the photon is created at the instant the atom radiates and is not present beforehand. Calculations made by Dirac (1927) on the scattering of photons by electrons suggested two steps in the scattering process: The incident photon is absorbed by the electron and the scattered photon is created. Fermi (1932) showed that the Coulomb force between charged particles could be thought of as an exchange interaction involving photons: One charged particle emitted a photon which was absorbed later by a second charged particle. It was not realized until 1934 that this type of description could also apply to other particles.

When Fermi (1934) developed his theory of β-decay he postulated that the electron and neutrino emitted in β-decay did not exist in the nucleus before the decay, but were created at the instant the decay took place, just as in the emission of light quanta. In this theory the total number of electrons and neutrons, like the number of photons in the theory of radiation, was not necessarily constant, since these particles could be created or destroyed in β-decay.

Heisenberg in his theory of nuclear structure suggested that the neutron–proton force was caused by an exchange of charge between nucleons (cf. Part 2, p. 145). He also hinted that the charge exchange might be somehow connected with the β-decay process, but he could not develop this idea any further at the time as there was no consistent theory of β-radioactivity. After Fermi had published his paper Tamm (1934) and Iwanenko (1934) investigated the β-theory of nuclear forces. They pointed out that the creation and subsequent annihilation of an electron and neutrino in the field of a proton and a neutron would lead to an exchange interaction between the nucleons, in the same way as the Coulomb interaction between two electrons was caused by the exchange of a photon between them. The magnitude of the exchange potential turned out to be

$$U(r) \simeq -g^2/\hbar c r^5$$

where g was Fermi's β-decay coupling constant. This coupling constant was very small, and the theoretical interaction was much weaker than required by empirical evidence from nuclear properties. Tamm and Iwanenko concluded, therefore, that the forces between protons and neutrons could not be related to β-decay. Although this idea was not successful it seems to have provided Yukawa with an important lead for his development of the meson theory of nuclear forces.

It is sometimes helpful to represent exchange processes by diagrams. They were introduced by Feynman† (1949) and have become part of the language of elementary particle physics. For example, Dirac's process for the scattering of photons by electrons can be represented by the diagrams in Fig. 10. In these diagrams a solid line represents an electron. It can be thought of as the trajectory of an electron in space–time, different points on the line indicating the position of the electron at various times. The arrows on the lines show progression from past to future, so that the initial state is to the left and the final state to the right in Fig. 10(a) and (b). Wavy lines denote photons. In Fig. 10(a) the incident photon is absorbed by the electron at a time t_0 and the scattered photon is emitted at a

† In Feynman's original paper the diagrams have a definite significance: Each one represents a particular term in the perturbation expansion of the interaction between the electromagnetic field and the charges. Subsequently diagrams have been used in many contexts, often in rather a loose way to give a physical picture of some interaction process. We use them here in this way.

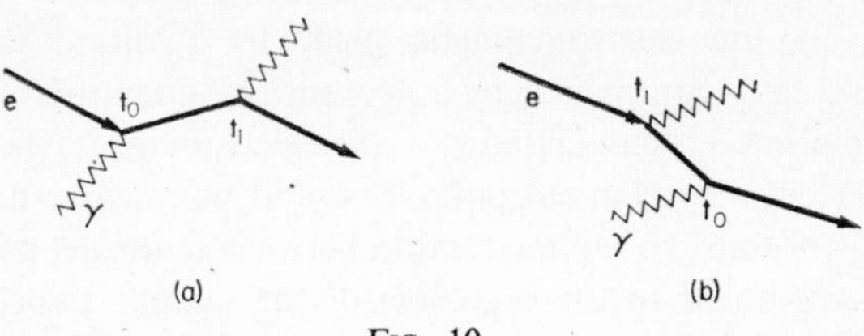

FIG. 10.

later time t_1. It is also possible that the incident photon is absorbed after the scattered photon is created, Fig. 10(b). Fig. 11(a) is a diagram of the interaction between charged particles due to photon exchange, and Fig. 11(b) represents the proton–neutron interaction as conceived by Tamm and Iwanenko.

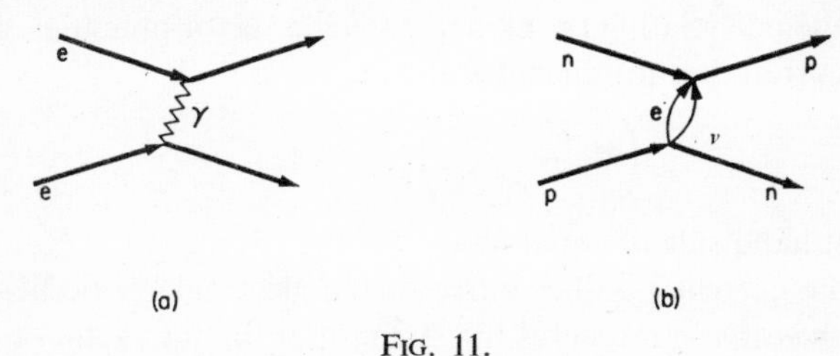

FIG. 11.

6.1 Yukawa's Theory

Yukawa (cf. Part 2, p. 214) took up the ideas of Heisenberg, Tamm and Iwanenko and extended them in a very definite way. The main defect of the β-theory of nuclear forces was that the neutron–proton interaction turned out to be much too small. In order to avoid this difficulty Yukawa proposed that a neutron could change into a proton by a process different from Fermi's β-decay mechanism and that the negative charge could be transferred to a nearby proton thus transforming it into a neutron. If this exchange process occurred with a large probability it would produce a strong interaction between the two nucleons. Up to this point Yukawa followed Heisenberg's discussion of exchange forces. He went on to suggest, however, that this interaction between nucleons could be described by a field of force just as the interaction between charged particles could be described by the electromagnetic field. In quantum theory this field should be accompanied by a new sort of quantum in the same way as the photon is associated with the electromagnetic field. As the interaction between charged particles could be attributed to an exchange of photons, so the interaction between neutrons and protons could be attributed to the exchange of the quanta associated with the new field. Such quanta are now called *mesons*.†

Yukawa assumed that the interaction could be represented by a scalar field U obeying the equation

$$\left(\nabla^2 - \frac{1}{c^2}\frac{\partial^2}{\partial t^2} - \lambda^2\right) U = -2\pi g\tilde{\psi}\tau^- \psi \tag{6.1}$$

where $\tilde{\psi}$ and ψ are nucleon wave functions. This equation is analogous to the one relating the electromagnetic vector potential $\mathbf{A}$ to the electric current $\mathbf{j}(r)$ producing the field

$$\left(\nabla^2 - \frac{1}{c^2}\frac{\partial^2}{\partial t^2}\right)\mathbf{A} = -4\pi\mathbf{j}$$

The right hand side of equation (6.1) is the source of the meson field just as the current $\mathbf{j}$ is the source of the electromagnetic field. The *coupling constant* g measures the strength of the interaction between the field U and the nucleons in the same way as the electronic charge e determines the interaction between electrons and an electromagnetic field. The isobaric spin operator τ^- appears in the equation because a nucleon changes from a proton state to a neutron state or vice versa when it interacts with the field U. Equation (6.1) determines the field produced by the nucleons. Yukawa supplemented it with an equation giving the force exerted by the field on the nucleons and he derived the neutron–proton exchange interaction

$$V = -\frac{g^2}{2}(\tau_1^\xi \tau_2^\xi + \tau_1^\eta \tau_2^\eta) \exp(-\lambda r)/r \tag{6.2}$$

This potential has exactly the same form as Heisenberg's exchange interaction with an exchange integral‡ $J(r) = g^2 \exp(-\lambda r)/r$. The

† The name "meson" was invented by H. J. Bhabha (1939).

‡ The sign of the exchange interaction is incorrect in Yukawa's original paper. He corrected it in a later publication (1937).

sign of the interaction is determined by the theory and cannot be chosen arbitrarily.

The quanta (mesons) accompanying Yukawa's field were predicted to have zero spin (because the field U was a scalar field) and obey Bose–Einstein statistics. The symmetry of the theory required that mesons with both positive and negative charge should exist with the same mass $m = \lambda\hbar/c$. This would have to be about 200 times larger than the electron mass in order to give the correct range for nuclear forces.

Yukawa's theory was not widely accepted at first, but in 1937 the μ-meson was discovered in cosmic rays. It had about the same mass as Yukawa had predicted, and its discovery awakened interest in the meson theory of nuclear forces. We know now that the μ-meson plays no part in the interaction between nucleons, but the belief that Yukawa's particle had been discovered sustained the efforts of physicists for the next decade, until the π-meson was found in 1947.

Early calculations of the meson exchange interaction used quantum mechanical perturbation theory. A Hamiltonian H was postulated which contained term H_{nuc} and H_{meson} describing free nucleons and mesons respectively and a term H' for the emission and absorption of mesons by nucleons,

$$H = H_{\text{nuc}} + H_{\text{meson}} + H'$$

The exchange interaction between nucleons can be calculated by treating the interaction H' between mesons and nucleons as a perturbation. To second order the interaction energy is

$$V_{NP} = -\sum_n \frac{H'_{in}H'_{nf}}{E_n - E_i} \tag{6.3}$$

The initial state i consists of a proton with position r_1 and a neutron at r_2. In the final state f there is a neutron at r_1 and a proton at r_2. The suffix n stands for all possible intermediate states in which a meson has been emitted by one of the particles. There are two groups of terms in equation (6.3) which are illustrated by the diagrams in Fig. 12. The intermediate state may contain a positive meson emitted by the proton and absorbed by the neutron (Fig. 12(a)) or a negative

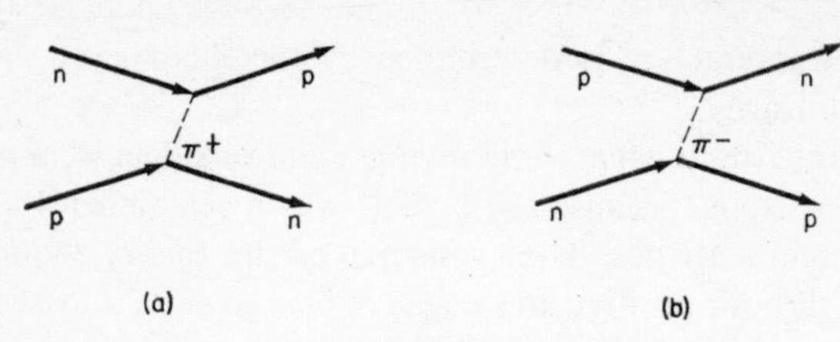

FIG. 12.

meson emitted by the neutron and absorbed by the proton (Fig. 12(b)). If the nucleons are assumed to be heavy and have fixed positions the energy difference between the initial and intermediate state $E_n - E_i$ is just the meson energy $\varepsilon = (\mathbf{p}^2 c^2 + m^2 c^4)^{\frac{1}{2}}$ where $\mathbf{p}$ is the meson momentum and m is its mass.

If the intermediate meson states are represented by plane waves $L^{-\frac{3}{2}} \exp(i\mathbf{k}.\mathbf{r})$ normalized in a volume $L^3(\hbar\mathbf{k} = \mathbf{p})$, then the matrix elements of the interaction H' can be calculated. We do not give a derivation here as it would involve a lengthy discussion of field theory. It can be found in many text books, e.g. Messiah, 1960, p. 985. The matrix elements H' are

$$\frac{gA\tau^-}{(2\varepsilon)^{\frac{1}{2}}} \exp(-i\mathbf{kr}) \quad \text{for a positive meson emitted by a proton} \tag{6.4}$$

$$-\frac{gA\tau^+}{(2\varepsilon)^{\frac{1}{2}}} \exp(-i\mathbf{kr}) \text{ for a negative meson emitted by a neutron}$$

The absorption matrix elements H_{nf} are just the complex conjugates of these. In these matrix elements A is a normalization constant $A^2 = 2\pi\hbar^2 c^2/L^3$ and g is the meson–nucleon coupling constant. Substituting in equation (6.3) we find

$$V_{NP} = -g^2 A^2 (\tau_1^+ \tau_2^- + \tau_2^+ \tau_1^-) \sum_n \exp(i\mathbf{k}_n \mathbf{r})/\varepsilon_n^2$$

where $\mathbf{r} = \mathbf{r}_1 - \mathbf{r}_2$. The sum over intermediate states can be calculated by converting it to an integral† and using the relation

$$\varepsilon^2 = \hbar^2 c^2 (k^2 + \lambda^2)$$

† $\Sigma_n F_n = \int F(k)\rho(k)\,\mathrm{d}\mathbf{k}$ where the density of states $\rho(k) = \left(\frac{L}{2\pi}\right)^3$

where $\lambda = mc/\hbar$,

$$A^2 \sum_n \exp(i\mathbf{k}_n.\mathbf{r})/\varepsilon_n^2 = \frac{L^3 A^2}{\hbar^2 c^2 (2\pi)^3} \int \frac{\exp(i\mathbf{k}.\mathbf{r})}{k^2+\lambda^2} \mathrm{d}\boldsymbol{k}$$

$$= \frac{1}{\pi r} \int_0^\infty \frac{k \sin kr}{k^2+\lambda^2} \mathrm{d}k$$

$$= \tfrac{1}{2}\exp(-\lambda r)/r$$

Hence we obtain Yukawa's formula

$$V_{NP} = -\frac{g^2}{2}(\tau_1^+ \tau_2^- + \tau_2^+ \tau_1^-) \exp(-\lambda r)/r \tag{6.5}$$

There can be no interaction between like nucleons in the second order perturbation approximation as there is no possibility for them to exchange a single charged meson. Two protons can interact by exchanging two positive mesons (Fig. 13), but the force due to the

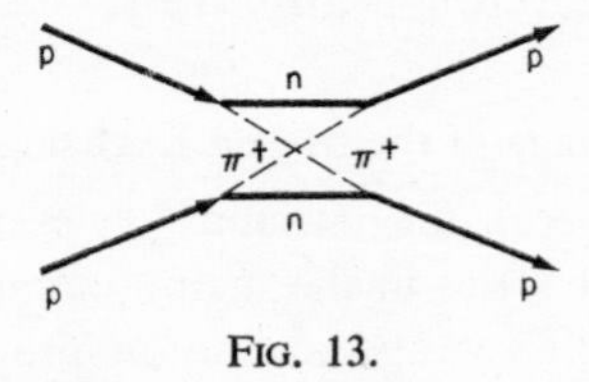

FIG. 13.

exchange of two particles must be different from the force caused by the exchange of one only. Thus Yukawa's original theory does not predict charge independent nuclear forces.

Kemmer (1938) and independently Yukawa and Sakata (1938) noticed that exchange forces could be made charge independent by introducing neutral mesons with the same mass as the charged meson. One neutral meson can be exchanged between any pair of nucleons. Provided the coupling constants are chosen correctly the nuclear force can be made the same between any pair of nucleons. The correct choice of coupling constants makes the matrix element for the emission of a neutral meson by a nucleon equal to

$$\frac{gA\tau^\zeta}{\varepsilon^{\frac{1}{2}}} \exp(-i\mathbf{k}.\mathbf{r})$$

Inclusion of neutral meson exchange modifies equation (6.5) for the exchange interaction in the following way

$$V = -g^2(\tau_1^\xi \tau_2^\xi + \tau_1^\eta \tau_2^\eta + \tau_1^\zeta \tau_2^\zeta) \exp(-\lambda r)/r$$
$$= -g^2(\boldsymbol{\tau}_1 . \boldsymbol{\tau}_2) \exp(-\lambda r)/r \qquad (6.6)$$

If this proposal is accepted, the isobaric spin classification can be extended to mesons. The positive, neutral and negative mesons in the theory are three states of an isobaric triplet ($T = 1$) with $T_0 = 1$, 0 and -1 respectively. Isobaric spin should be conserved not only in reactions involving nucleons, but also in reactions involving nucleons and mesons.

In equation (6.6) $\boldsymbol{\tau}_1 . \boldsymbol{\tau}_2 = 1$, if the two nucleons have total isobaric spin $T = 1$ and $\boldsymbol{\tau}_1 . \boldsymbol{\tau}_2 = -3$ if $T = 0$. The exchange interaction is charge independent, but it is attractive in $T = 1$ states and repulsive in $T = 0$ states. Empirically, the neutron–proton interaction must be attractive in $T = 0$ states in order to bind the deuteron, hence the exchange potential (6.6) is not satisfactory.

6.2 The Range of the Meson Exchange Potential

In perturbation theory the internucleon exchange interaction is expanded as a power series in the meson coupling constant g. The second order theory gives a contribution proportional to g^2, the fourth order one proportional to g^4 and so on. In the Feynman diagram representation there is a factor g for every meson emitted or absorbed by a nucleon. Thus second order perturbation theory means one meson exchange (Fig. 12), fourth order theory means two meson exchange (Fig. 13). All early calculations used second order perturbation theory, but already in 1938 the shortcomings of this method were recognized. Fröhlich, Heitler and Kemmer (1938) calculated the second and fourth order exchange energies and found that although the fourth order interaction had a shorter range than the second order, it was also much larger for small separations of the interacting nucleons. It seemed that the perturbation series would converge very slowly, if the two nucleons were close together. If the perturbation method was so unreliable, could any of its predictions be trusted?

One of the important predictions of Yukawa's theory is the relation between the range ρ of the nuclear forces and the mass m of the meson responsible for the exchange interaction

$$\rho \simeq \hbar/mc$$

Wick (1938) was able to show that this prediction was not dependent on perturbation theory but could be derived simply by a general argument based on Heisenberg's Uncertainty Principle. We reproduce Wick's paper in Part 2 of this book.

His argument shows at once that a two meson exchange process such as the one illustrated in Fig. 13 must have a range about half as large as that of a one meson exchange interaction. If several different sorts of mesons contribute, then the long range part of the interaction will be due mainly to exchange of the lightest mesons.

Later it was recognized that even the radial dependence

$$\exp(-\lambda r)/r$$

of Yukawa's interaction for large enough $r(\lambda r \gtrsim 1)$ did not depend on details of his theory and would be a consequence of any exchange theory. To illustrate this point we calculate the exchange interaction between two heavy particles using a quite different model.

In his original paper Heisenberg suggested the nuclear force might be due to a resonance exchange mechanism similar to that binding an H-atom and H^+ ion in the hydrogen molecular ion. We shall calculate the exchange force predicted by such a model.

Let us consider a light particle P with mass m interacting with two heavy particles A and B at fixed positions r_A and r_B. The interaction of the light particle with A and B is represented by two similar potentials $V_A = V(\mathbf{r}-\mathbf{r}_A)$ and $V_B = V(\mathbf{r}-\mathbf{r}_B)$. If the heavy particles A and B are near each other then the interaction of each of them with the light particle will produce an exchange interaction between them by the resonance exchange effect. We shall calculate the exchange energy $J(\mathbf{r}_A-\mathbf{r}_B)$ (cf. Chap. 2.2) assuming that the potential $V(r)$ has a short range ($V(\mathbf{r}) = 0$, if $r > b$), that the light particle has a weakly bound state with binding energy ε in the potential $V(\mathbf{r})$ and that the separation $R = |\mathbf{r}_A-\mathbf{r}_B|$ between A and B is much larger than the range of the force b.

The wave functions ϕ_A and ϕ_B introduced in Chap. II are

$$\phi_A = \phi(\mathbf{r}-\mathbf{r}_A) \quad \text{and } \phi_B = \phi(\mathbf{r}-\mathbf{r}_B)$$

where $\phi(\mathbf{r})$ is the wave function of the bound state of the light particle in the potential $V(\mathbf{r})$. If $r > b$ the wave function $\phi(\mathbf{r})$ satisfies the Schrödinger equation

$$\nabla^2 \phi = \gamma^2 \phi \tag{6.7}$$

where $\gamma = (2m\varepsilon)^{\frac{1}{2}}/\hbar$. Hence for $r > b$ the wave function is

$$\phi(r) = A \exp(-\gamma r)/r \tag{6.8}$$

where A is a normalization constant chosen so that $\int_0^\infty \phi^2 d\mathbf{r} = 1$. If we put $A = N(\gamma/2\pi)^{\frac{1}{2}}$ then N is dimensionless with a value approximately equal to 1. The exchange energy is (cf. equation 2.3)

$$J = \int \phi_A V_B \phi_B d\mathbf{r}$$

To evaluate J we remember that V_B has a short range, so the integrand in this equation vanishes unless $|\mathbf{r}-\mathbf{r}_B| < b$. As $R \gg b$, ϕ_A is approximately constant over this region and equal to $\phi(R)$. Hence

$$J \simeq \phi(R) \int V_B \phi_B dr \tag{6.9}$$

The integral in equation (6.9) can be transformed using the Schrödinger equation for ϕ_B, $(T+V_B)\phi_B = -\varepsilon\phi_B$ giving

$$\int V_B \phi_B d\mathbf{r} = -\varepsilon \int \phi_B d\mathbf{r} + \frac{\hbar^2}{2m} \int \nabla^2 \phi_B d\mathbf{r} \tag{6.10}$$

where both integrals are taken over all space. The second integral is identically zero (it can be transformed to a surface integral $\int_S \nabla\phi_B . d\mathbf{S}$ over a large surface S and $\nabla\phi_B$ vanishes exponentially for large r). If $\gamma b \ll 1$ the main contribution to the integral of the wave function comes from the region $r > b$, so $\phi(\mathbf{r})$ can be approximated by its asymptotic form given in equation (6.8). Thus

$$\int \phi(\mathbf{r}) d\mathbf{r} \simeq 4\pi A/\gamma^2$$

Collecting these results we find that

$$\begin{aligned} J(r) &= -\varepsilon 4\pi A^2 \exp(-\gamma R)/\gamma^2 R \\ &= -2\varepsilon N^2 \exp(-\gamma R)/\gamma R \end{aligned} \tag{6.11}$$

This result has a number of interesting features:

(i) The exchange energy has the same dependence on the interparticle separation R as Yukawa's potential.

(ii) The calculation makes approximations, but they are quite different from those used in meson theory. In particular, there is no perturbation expansion as a power series in the coupling constant ($g^2 \sim 2\varepsilon N^2$).

(iii) The exchange interaction depends only on the asymptotic properties of the wave function ϕ, i.e. on its binding energy ε and normalization N. Details of the potential $V(r)$ do not enter into the result. This fact suggests that equation (6.11) should be true, even if A and B are complex particles. Suppose a complex particle A can break up into a particle B emitting a light particle P. The process requires energy and has a threshold at an energy ε. If we look at the particle A, we might sometimes find that it consists of B together with the light particle P separated by some distance.† The normalization constant N measures the probability of this result. In other words, the wave function of A contains a component which corresponds to its separation into $B+P$, and N is the normalization of this component. In the language of nuclear physics N is related to the "reduced width" for the process $A \to B+P$. In general $N^2 \lesssim 1$.

(iv) We derived the formula (6.11) assuming the light particle to be non-relativistic, but it remains true even if the particle P has to be treated relativistically, provided the correct relation between the binding energy ε and the constant γ is used. In relativistic mechanics the threshold energy ε required to break up A into $B+P$ is

$$\varepsilon = mc^2 + M_B c^2 - M_A c^2$$

The Schrödinger equation (6.6) must be replaced by the Klein–Gordon equation

$$(\mathbf{p}^2 c^2 + m^2 c^4)\phi = E^2 \phi \tag{6.12}$$

where $\mathbf{p} = -i\hbar\nabla$ and $E = -\varepsilon + mc^2$. Comparing equations (6.12) and (6.6) we find that

$$\hbar^2 c^2 \gamma^2 = m^2 c^4 - (mc^2 - \varepsilon)^2 \tag{6.13}$$

† This is not inconsistent with the principle of energy conservation because an observation can change the energy of a system.

Thus, if we agree that a proton sometimes looks like a neutron plus a positive meson in an S-state†, the Heitler–London resonance effect predicts an exchange interaction of the Yukawa type. The proton has almost the same mass as the neutron, so $\varepsilon = mc^2$ and from equation (6.13)

$$\gamma = mc/\hbar$$

The range of the force is the same as in Yukawa's theory which could be expected from Wick's argument.

These considerations suggest that some of the predictions of Yukawa's theory are very general. *Any* meson exchange theory will relate the range of the exchange interaction to the meson mass in the same way and will give the Yukawa radial dependence for the exchange potential, provided the nucleons are not too close together.

Meson theory has suffered many modifications since Yukawa published his paper, but these two predictions have survived. It is important to remember that the formula (6.6) or (6.11) for the radial dependence of the interaction holds only for internucleon separations large enough for two meson exchange to be neglected. If the separation is too small then several mesons may be exchanged and the resulting force will be very complicated.

6.3 Pseudo-Scalar and Vector Mesons

In Section 6.1 we saw that some predictions of Yukawa's original theory were not consistent with experimental facts. Yukawa's exchange interaction could not bind the deuteron and was not charge independent. A later paper (Yukawa and Sakata, 1938) pointed out that his original theory had many simple variations, so that it might be possible to find an acceptable one amongst them. They made an extensive investigation of the vector meson theory. At the end of Section 6.1 we showed how Kemmer (1938) and Yukawa (1938) were able to make Yukawa's original theory charge independent by postulating the existence of a neutral meson with the same mass as

† It is often called a "virtual" meson. The proton can emit a meson, but this process does not conserve energy, so it must be absorbed again within a time $\Delta t < \hbar/mc^2$ given by Heisenberg's Uncertainty Principle.

the charged mesons and with appropriate coupling constants to protons and neutrons. At about the same time they also made a systematic investigation of the possible variations of Yukawa's theory and showed how the characteristics of an exchange interaction depended on the intrinsic spin and parity of the mesons.

The mesons in Yukawa's theory have zero spin, and their coupling to nucleons is such that a nucleon can emit or absorb a meson in an S-state, i.e. with orbital angular momentum $l=0$. As the parity of an S-state is positive the meson must have *positive intrinsic parity* if the meson–nucleon interaction is to conserve parity. There is an alternative coupling for zero spin mesons, namely a meson may be emitted or absorbed in a P-state (orbital angular momentum $l=1$). The parity of a P-state is negative, hence the meson must have *negative intrinsic parity* if the theory is to conserve parity. The law of conservation of parity predicts selection rules in reactions involving mesons, and the intrinsic parity of a meson may be determined by studying such reactions. We shall give an example in Section 6.5. Zero spin mesons with positive intrinsic parity are called *scalar* mesons. Those with negative intrinsic parity are called *pseudoscalar* mesons. Kemmer also discussed *vector* mesons (spin 1, negative parity) and *pseudovector* mesons (spin 1, positive parity). These names have their origin in field theory and refer to the transformation properties of the fields associated with mesons with respect to rotation and reflection of coordinate axes. We summarize Kemmer's (1938a) results for these different mesons. (Kemmer's potentials are calculated in the "static approximation" assuming the nucleons to be heavy with fixed positions.)

(i) *Scalar mesons* (spin and parity $J\pi=0+$): This is Yukawa's original theory. The meson field surrounding the nucleon is in an S-state and is spherically symmetrical. The exchange interaction is central and spin independent. If the theory contains both charged and neutral mesons with couplings to give charge independence, then the exchange potential is

$$V(r) = -\frac{f^2}{4\pi}\hbar c \boldsymbol{\tau}_1 . \boldsymbol{\tau}_2 \exp(-\mu r)/r \qquad (6.14)$$

where $\mu = mc/\hbar$ and m is the meson mass. The coupling constant f is dimensionless.

(ii) *Pseudoscalar Mesons* ($J\pi = 0-$): A virtual meson in the field of a nucleon is in a P-state. This state is not spherically symmetrical, and its orientation in space is determined by the spin of the nucleon. The interaction between two nucleons depends on the relative orientation of their meson fields, hence it is spin dependent and non-central. The perturbation matrix element H'_{in} of the meson–nucleon interaction (cf. Section 6.1) contains a factor $\boldsymbol{\sigma}.\mathbf{k}$, where $\boldsymbol{\sigma}$ is the Pauli-spin matrix of the nucleon and $\hbar\mathbf{k}$ is the momentum of the meson in the intermediate state n. The interaction potential is

$$V(r) = \tfrac{1}{3}\frac{f^2}{4\pi}\hbar c\boldsymbol{\tau}_1.\boldsymbol{\tau}_2\left[\boldsymbol{\sigma}_1.\boldsymbol{\sigma}_2 + S_{12}\left(1 + \frac{3}{\mu r} + \frac{3}{(\mu r)^2}\right)\right]\exp(-\mu r)/r \tag{6.15}$$

$S_{12} = 3(\boldsymbol{\sigma}_1.\mathbf{r})(\boldsymbol{\sigma}_2.\mathbf{r})/r^2 - \boldsymbol{\sigma}_1.\boldsymbol{\sigma}_2$ is the tensor force operator (cf. Chap. 3).

(iii) *Vector Mesons* ($J\pi = 1-$): A vector meson emitted or absorbed by a nucleon is in a P-state, but there are two independent forms of coupling. The orbital ($l = 1$) and spin ($J = 1$) angular momentum of the meson can couple to a resultant of either zero or one, and this can then couple with the nucleon spin to give a total angular momentum $\frac{1}{2}$. In the first case the matrix element of the meson–nucleon interaction contains a factor $\mathbf{J}.\mathbf{k}$ ($\mathbf{J}$ is the meson spin vector) and in the second case a factor $\boldsymbol{\sigma}.\mathbf{J}\times\mathbf{k}$. In other words, the meson spin is polarized parallel to its direction of motion (longitudinal polarization) in the first case and perpendicular to its direction of motion (transverse polarization) in the second case. Each of the two possibilities has its own coupling constant. The longitudinal coupling ($\mathbf{J}.\mathbf{k}$) does not involve the nuclear spin, hence its contribution V_l to the internucleon interaction is spin independent. The transverse coupling ($\boldsymbol{\sigma}.\mathbf{J}\times\mathbf{k}$) gives rise to a non-central spin dependent interaction Vt. The total interaction is

$$V = V_l + V_t$$

where

$$V_l = \frac{f_l^2}{4\pi} \hbar c \boldsymbol{\tau}_1 . \boldsymbol{\tau}_2 \exp(-\mu r)/r \tag{6.16}$$

and

$$V_t = \frac{f_t^2}{4\pi} \hbar c \boldsymbol{\tau}_1 . \boldsymbol{\tau}_2 \left[\tfrac{2}{3} \boldsymbol{\sigma}_1 . \boldsymbol{\sigma}_2 - \tfrac{1}{3} S_{12} \left(1 + \frac{3}{\mu r} + \frac{3}{(\mu r)^2} \right) \right] \exp(-\mu r)/r \tag{6.17}$$

If the nucleons are at rest, a meson emitted with its spin polarized parallel (or perpendicular) to its direction of motion is still polarized in this way when it is absorbed. It is for this reason that the two couplings do not mix. The situation is different if the nucleons are moving relative to one another. On changing from one frame of reference to another moving relative to it, angular and linear momentum transform in a different way, because, relativistically, momentum is a part of a 4-vector and angular momentum is a part of an antisymmetric tensor. Thus, if a meson is emitted by one nucleon with its spin polarized parallel to its direction of motion, it will have a small component of transverse polarization as seen from the second nucleon. The meson can be emitted by the longitudinal and absorbed by the transverse coupling. This effect produces a spin orbit force (Breit 1960),

$$V_{so} = \frac{4\hbar c}{\mu^2} f_t f_l \left(\frac{m}{M} \right) \mathbf{L} . \mathbf{S} \frac{1}{r} \frac{\mathrm{d}}{\mathrm{d}r} \exp(-\mu r)/r \tag{6.18}$$

where M is the nucleon mass and m the meson mass. If we replace $\boldsymbol{\tau}_1 . \boldsymbol{\tau}_2$ by one, put

$$f_l^2 = 4\pi \frac{e^2}{\hbar c}, \quad f_t^2 = 4\pi \frac{e^2}{\hbar c} \left(\frac{m}{2M} \right)^2 g_p^2$$

and let $\mu \to 0$, then equation (6.16) becomes equal to the Coulomb force between two protons, equation (6.17) is the interaction between their magnetic moments and equation (6.18) is the electromagnetic spin orbit coupling. Photons are neutral vector mesons with zero mass. (g_p is the proton g-factor.)

(iv) *Pseudo vector Mesons* ($J\pi = 1+$): Again there is a longitudinal and transverse coupling. The potential is

$$V = -\tfrac{1}{3}\hbar c \boldsymbol{\tau}_1 . \boldsymbol{\tau}_2 \times$$
$$\times\left[(f_l^2+2f_t^2)\boldsymbol{\sigma}_1 . \boldsymbol{\sigma}_2+(f_l^2-f_t^2)S_{12}\left(1+\frac{3}{\mu r}+\frac{3}{(\mu r)^2}\right)\right]\times$$
$$\times \exp(-\mu r)/r \qquad (6.19)$$

Kemmer's mesons had isobaric spin $T=1$ and could have either positive, negative or zero charge. Bethe (1939) suggested another way to make a meson theory charge independent. He postulated that perhaps only neutral mesons were exchanged and the meson coupling strength was the same with both protons and neutrons. His mesons had isobaric spin $T=0$. They could be of scalar, pseudoscalar, vector or pseudovector type, and the exchange potentials could be obtained from those for the charged meson theory by replacing $\boldsymbol{\tau}_1 . \boldsymbol{\tau}_2$ by unity. In order to see the predictions of these potentials for the nucleon–nucleon interaction at low energies we give (Table 5) the central and tensor parts of the eight possible

TABLE 5

Meson Potentials in Even Parity States

Type of Meson		Singlet Even	Triplet Even
Scalar	$T=0$	$-V_c$	$-V_c$
	$T=1$	$-V_c$	$3V_c$
Pseudoscalar	$T=0$	$-V_c$	$\frac{1}{3}(V_c+V_T)$
	$T=1$	$-V_c$	$-V_c-V_T$
Vector	$T=0$	$(1-2k^2)V_c$	$(1+\frac{2}{3}k^2)V_c-\frac{1}{3}V_T$
	$T=1$	$(1-2k^2)V_c$	$-(3+2k^2)V_c+V_T$
Pseudovector	$T=0$	$(1+2k^2)V_c$	$-\frac{1}{3}(1+2k^2)V_c-\frac{1}{3}(1-k^2)V_T$
	$T=1$	$(1+2k^2)V_c$	$(1+2k^2)V_c+(1-k^2)V_T$

$V_c = \frac{f^2}{4\pi}\hbar c \exp(-\mu r)/r \qquad V_T = \frac{f^2}{4\pi}\hbar c\, S_{12}\left[1+\frac{3}{\mu r}+\frac{3}{\mu^2 r^2}\right]\exp(-\mu r)/r$

$k^2 = f_t^2/f_l^2$

potentials in singlet even states (states with even orbital angular momentum $S=0$, $T=1$, $\boldsymbol{\sigma}_1.\boldsymbol{\sigma}_2=-3$, $\boldsymbol{\tau}_1.\boldsymbol{\tau}_2=1$) and triplet even states ($T=0$, $S=1$, $\boldsymbol{\tau}_1.\boldsymbol{\tau}_2=-3$, $\boldsymbol{\sigma}_1.\boldsymbol{\sigma}_2=1$).

Kemmer asked the following question: Is one of the single meson exchange potentials consistent with the experimental deuteron binding energy and low energy nucleon–nucleon scattering? (A condition was that the force must be attractive in both singlet even and triplet even states and rather stronger in the triplet state.) He considered only the charged ($T=1$) meson theories and did not realize that the tensor force could contribute to the deuteron binding energy. The scalar and pseudovector theories could be excluded immediately, because they were repulsive in either singlet or triplet states. Kemmer favoured the vector theory, since the two coupling constants f_l and f_t could be adjusted (if $f_t > f_l$) to give the correct interaction strengths in both spin states.

In 1939 Bethe introduced his neutral meson theory. In the meantime the deuteron quadrupole moment had been measured and was known to have a positive sign. Bethe (1939) showed that the tensor forces in the charged vector theory would give the wrong sign for the deuteron quadrupole moment and would make the deuteron charge distribution "saucer" shaped instead of "cigar" shaped as indicated by the measurements of Kellogg *et al.* (1939). Bethe himself was in favour of the neutral vector theory. The central part of the triplet even potential is repulsive, but the tensor force is strong enough to overcome the central repulsion and bind the deuteron. Bethe made some numerical calculations and showed that this theory would give about the correct value for the deuteron quadrupole moment. He also pointed out a difficulty with the tensor potential: It had a $1/r^3$ singularity at the origin and the Schrödinger equation has no solution if the potential contains such a strong divergence. It was known, however, that the one-meson exchange potential was unreliable for small separations of the interacting nucleons. In order to make numerical calculations Bethe cut off the singularity in an arbitrary way by making the strength of the tensor potential constant for radii smaller than some critical radius. His results were not sensitive to the way in which the cut-off was made. Thus far Bethe's theory seemed very successful although it contained one undesirable feature. It required neutral mesons whereas charged mesons had been discovered in cosmic rays and it was this fact that had made the meson theory of nuclear forces credible.

Looking at Table 5 we see that the charged pseudoscalar theory also has many desirable aspects. The central potential is attractive in both singlet and triplet states, and the sign of the tensor force is consistent with a positive deuteron quadrupole moment. By 1940 the theory had gone about as far as it could and more experimental information about the properties of mesons was needed to guide further efforts. Many people still believed that all properties of matter could be accounted for by a few elementary particles and thought that Nature had been generous in supplying one meson to explain nuclear forces. Now it seems that she preferred to use many variations and at least four of the meson varieties discussed in this section seem to play an important role in nuclear forces (cf. Section 6.8).

6.4 The π-Meson

When Anderson and Neddermeyer (1937) discovered the μ-meson in cosmic rays, most physicists believed that Yukawa's prediction had been verified and that this meson must be responsible for nuclear forces. The μ-meson had about the right mass. The coupling constant g could be predicted from the strength of the nuclear forces. With this information cross sections for the scattering of mesons could be estimated, but the calculated values were always much too large to be consistent with cosmic ray observations. A great deal of theoretical effort was directed to explaining this discrepancy, but only rarely was the suspicion expressed that there might be several types of mesons (Sakata and Inoue 1946). In 1947 the stage was set for another revolution.

Conversi, Pancini and Piccioni (1947), of Rome University, had conducted an experiment to investigate the interactions of slow negative mesons with matter. Theoretical calculations by Tomonaga and Araki (1940) had suggested that a positive meson passing through a dense material would lose its energy by collisions with atomic electrons and ultimately come to rest. When its energy fell below a certain value it could no longer approach an atomic nucleus in the material because of Coulomb repulsion and would finally decay into a positron plus neutrinos with its natural lifetime of about 10^{-6} sec. A negative meson, on the other hand, could come to rest in the same way as a positive meson, but it would then be attracted towards the positively charged nucleus of an atom of the material by electrostatic forces. If, according to Yukawa's predictions, the meson interacted very strongly with nucleons, it would be captured by a nucleus quickly (within 10^{-11} sec) causing a nuclear disintegration. The lifetime for capture would be shorter than the lifetime for a natural μ-decay by a factor of 10^5, hence almost all negative mesons should cause disintegrations. Conversi and his collaborators confirmed this prediction for mesons coming to rest in iron, but they found that an appreciable fraction of the negative mesons decayed by β-decay, if graphite was used for stopping the mesons. There was a striking discrepancy between the experimental result and the theoretical prediction of Tomonaga and Araki (1940). Fermi, Teller and Weisskopf (1947) and, independently, Wheeler (1947) investigated the theoretical mechanism for capturing mesons in more detail and confirmed that the conclusions of Tomonaga and Araki were correct. We reprint the paper of Fermi *et al.* in Part 2 of this book. They found that a slow negative meson would be captured into a Bohr *K*-orbit around an atomic nucleus within 10^{-12} sec. Once in this orbit, it should be captured by the nucleus within 10^{-18} sec. The experiments of Conversi *et al.* showed, however, that capture took place only after about 10^{-6} sec. There was a discrepancy of a factor of 10^{12} between the theoretical prediction and the experimental result proving conclusively that the μ-meson found in cosmic rays could not be the cause of nuclear forces.

The theoretical discussion of Fermi *et al.* was published in February 1947 and in May a new meson was discovered. A new method had been developed for studying cosmic radiation by photographic plates. It was shown by Perkins (1947) and by Occhialini and Powell (1947) that the tracks of charged mesons brought to rest in the photographic emulsion could be detected, if such plates were exposed to cosmic radiation at mountain altitudes. The meson masses could be estimated by grain counts and by studying deviations in the trajectories due to multiple Coulomb scattering. They were found to be about two hundred electron masses. It seemed that at least some of these particles were identical with the μ-mesons of the penetrating component of cosmic radiation. In May 1947, Lattes, Muirhead, Occhialini and Powell, of Bristol University, published two photographs each showing a meson stopping in the photographic emulsion and a secondary meson starting at the same point with a kinetic energy of about 2 MeV. The authors suggested that each secondary meson track was due to the decay of a heavy meson into a light one. Grain counts established the masses of both mesons to be about 100 MeV. If the heavy meson decayed into a light one emitting only one recoil particle (a neutrino or photon), the 2 MeV recoil energy of the light meson would imply a mass difference of about 25 MeV. The light meson was identified with the well-known μ-meson and the new heavier particle was called the π-meson, or pion.

In June 1947 an important conference on the Foundations of Quantum Mechanics was held at Shelter Island, U.S.A. The news that Powell and his coworkers had discovered the π-meson, had not reached America at the beginning of the conference, and the problems arising from the experiments of the Rome group were the main subject of discussion. Bethe and Marshak (1947) took up the *two meson hypothesis* suggested earlier by Sakata and Inoue (1946). They showed that many of the experimental anomalies in cosmic ray physics could be explained if two mesons with different masses exised in nature: A heavy meson supposed to be produced with a large cross section in the upper atmosphere and to be identified with Yukawa's particle that was responsible for nuclear forces; and the lighter μ-meson which was to be regarded as a decay product of the heavy one. As soon as the discovery of the Bristol group became known the whole pattern fell into place, and a new stage in the development of nuclear physics had begun.

In 1932 the discovery of the neutron coincided with important developments in the construction of particle accelerators (cf. Chap. I). A similar situation occurred in 1947. Energies attainable with conventional cyclotrons were limited to about 40 MeV. Such an accelerator with a 184 in. magnet was being constructed at the Berkeley Radiation Laboratory in 1945 under the direction of Lawrence. Just at that time the principle of "phase stability" was discovered independently by Veksler (1945) of the U.S.S.R. and by McMillan (1945) of the Berkeley Radiation Laboratory. The Berkeley group decided immediately to convert their 184 in. cyclotron into a synchro-cyclotron. It was operating within a year, i.e. in November 1946 and could produce deuterons with an energy of 190 MeV and helium ions with an energy of 380 MeV. π-mesons were produced by this accelerator early in 1948 (Gardner and Lattes 1948). The π-meson intensity was greater than that in cosmic rays by a factor of 10^8, and soon the main properties of the π-meson were established using this and similar accelerators constructed in other laboratories.

In 1950 the neutral π-meson (π^0) was found confirming the prediction of Kemmer (1938), Yukawa and Sakata (1938) and Fröhlich, Heitler and Kemmer (1938) based on the charge independence hypothesis. A number of experiments established the spin of both charged and neutral π-mesons to be zero, and in 1951 the parity of the π-meson was found to be negative (Panofsky 1951). This last measurement involved the capture of negative π-mesons by deuterons

$$\pi^- + D \rightarrow 2n$$

If the mesons have zero spin and are captured from a K-orbit (cf. Fermi *et al.*, Part 2, p. 227), the total angular momentum of the initial state must be $J=1$, equal to the deuteron spin. The states of two neutrons produced in the reaction are restricted by the Pauli exclusion principle and only the $^3P_1$ state has $J=1$ (cf. p. 78). This state has negative parity. The orbital parity in the initial state is positive so the capture reaction is allowed by parity conservation only if the meson has *negative intrinsic parity*. The reaction was observed and so the π-meson parity was negative. These results showed that the π-meson was a *pseudoscalar* particle.

6.5 The Pion–Nucleon Coupling Constant

After the discovery of the pion it seemed very likely that this particular type of meson was responsible for nuclear forces. Before the theory could be tested quantitatively it was necessary to know the value of the pion–nucleon coupling constant f. In the original theory of Yukawa this constant was a coefficient determining the magnitude of the interaction term in the Hamiltonian of the meson–nucleon system. Consequences of the theory were calculated by perturbation theory, i.e. by expanding observable quantities as a power series in the coupling constant. In early work only the lowest order perturbations were calculated, but one soon realized they were inadequate. Calculations of higher terms in the perturbation series encountered difficulties, because many of the integrals occurring in them diverged. At this stage in the development of field theory the physical interpretation of the coupling constant was not at all clear.

The same difficulties also existed in the theory of the interaction of electromagnetic radiation with electrons. In 1947 the technique of "renormalizing" a field theory was developed and applied successfully to calculations of the Lamb shift and the anomalous magnetic moment of the electron. This method distinguished between the unrenormalized electric charge e_0, a theoretical quantity which was a coefficient in the interaction term of the Hamiltonian of the radiation and electron fields, and the renormalized electric charge e, a physical quantity which could be measured experimentally. These two quantities had different numerical values and only the renormalized quantity e had any physical significance. The renormalization method was able to calculate finite results from a divergent theory by suppressing the unobservable unrenormalized quantities (which were infinite) and expressing all results in terms of finite renormalized quantities.

The situation was somewhat similar in meson theory, but it was not until 1954 that the definition of the renormalized coupling constant was clarified. Kroll and Ruderman (1954) suggested that the renormalized coupling constant should be defined so that its value could be determined directly from some experiment. They pointed out that the charge of an electron could be defined by the low energy limit of the cross section for scattering of photons by electrons (Thomson scattering)

$$\sigma_{\text{Thomson}} = \frac{8\pi}{3} \frac{e^2}{m^2 c^4} \tag{6.20}$$

and proposed by analogy that the low energy limit of the cross section for photo-production of charged π-mesons

$$\gamma + p \rightarrow n + \pi^+$$

could be used to define the meson–nucleon coupling constant. The idea was based on the following theorem proved by Kroll and Rudermann: The matrix element for charged photo-meson production at threshold correct to all orders in the meson coupling constant approaches the result calculated by second order perturbation theory in the limit of vanishing meson mass, provided the meson coupling constant and the nucleon mass are replaced by their renormalized values. Stated in another way, the theorem says that near threshold the cross section for photo-production of charged mesons should be given by the formula†

$$\sigma(\gamma, \pi^{\pm}) = 2e^2 f^2 k\mu^{-3} \tag{6.21}$$

where k is the meson wave number and $\mu = mc/\hbar$, m being the pion mass. Thus a measurement of the cross section near threshold could determine the renormalized pion–nucleon coupling constant f. The first measurements were made by Bernardini and Goldwasser in 1955 who found a value for the coupling constant

$$f^2/4\pi = 0{\cdot}073 \pm 0{\cdot}007$$

In order to correct for nucleon recoil effects it was necessary to investigate both the $p(\gamma, \pi^+)n$ and the $n(\gamma, \pi^-)p$ reactions. Information about the second reaction was obtained by studying the $\gamma + H^2 \rightarrow 2p + \pi^-$ reaction.

In 1956 Chew and Low applied the pseudoscalar meson theory to a discussion of the scattering of pions by nucleons. They assumed

† There are some corrections due to the recoil of the nucleon not included in equation (6.21). These depend on the ratio meson mass/nucleon mass.

that the nucleon was very heavy, so that its recoil could be neglected (the so-called static model) but calculated the scattering without using perturbation theory. These calculations fitted the observed pion–nucleon scattering data for pion energies up to about 300 MeV with just two parameters, the coupling constant f and a high energy cut-off parameter. Chew and Low also showed that the same coupling constant should describe both the pion–nucleon scattering and the photo production process. The pion coupling constant determined from meson scattering experiments was

$$f^2/4\pi = 0{\cdot}08$$

and in good agreement with the value found from the photo production experiment.

The wave function of the nucleon with its meson cloud can in principle be written down explicitly in the static model of Chew and Low. If $|\mathrm{p}, \alpha)$ is the complete wave function for a proton with its centre at the origin and with spin quantum number α, and $|\mathrm{n}, \beta; \mathbf{r})$ is the wave function of a neutron at the same point with spin quantum number β together with a π-meson at the point $\mathbf{r}$, these two wave functions have a finite overlap

$$(p, \alpha|n, \beta; \mathbf{r}) \simeq \frac{f\mu^{-\frac{1}{2}}}{4\pi} (\alpha|\boldsymbol{\sigma}|\beta).\,\mathrm{grad}\,\exp(-\mu r)/r \tag{6.22}$$

if r is large enough. In other words, the wave function for a proton contains a component that represents a neutron and a π-meson in a P-state with wave function proportional to grad $\exp(-\mu r)/r$. The coupling constant f determines the normalization of this component. In the (γ, π^+) reaction the γ-ray excites this "virtual" π-meson into an unbound state. The cross section (6.21) for the process may be calculated from the "wave function" (6.22) and the usual quantum mechanical formula for the photoelectric effect.

If the "virtual" meson clouds of two neighbouring nuclei overlap there will be an exchange interaction between them as discussed in Section 6.2. If they are far enough apart, so that only one-meson exchange is important, the exchange interaction is determined by the normalization of the virtual meson state, i.e. by the renormalized coupling constant f. Alternatively, the interaction energy is determined by second order perturbation theory (equation 6.15) provided the renormalized coupling constant is used.

Soon after the static model had been studied by Chew and Low the method of dispersion relations was introduced into elementary particle physics. This method has some features of traditional quantum mechanics; in several important respects, however, it is different. To explain the differences, we shall consider a typical problem in classical quantum mechanics, i.e. the theory of atomic structure.

An atom is assumed to consist of a number of electrons moving around a central nucleus. The electrons interact with each other and with the nucleus by electrostatic forces, and their motion is treated by the methods of quantum mechanics. The problem of atomic structure retains many aspects of a problem in classical mechanics. The forces are derived from classical mechanics and even the equations of motion look the same when written in the Heisenberg form. New physical characteristics introduced by quantum mechanics are Heisenberg's uncertainty principle, the Pauli exclusion principle and spin angular momentum. In a similar way the quantum theory of the electromagnetic field is just Maxwell's classical theory modified to make it consistent with the principles of quantum mechanics. Thus, most problems in traditional quantum mechanics are investigated by taking a theory of classical mechanics and modifying it according to definite rules in order to introduce specifically quantum mechanical features. In almost all cases the forces acting on the components of the quantum system are identical with the forces in the corresponding classical system. This fact made the Bohr correspondence principle such a powerful guide in the early developments of quantum mechanics.

In 1943 Heisenberg showed that the result of a general scattering process was completely determined by the scattering matrix. This matrix related the state of the interacting particles a long time after a scattering to their state a long time before the scattering. Only the final result of a scattering process could be observed. Heisenberg suggested a theory should not attempt to describe the unobservable intermediate stages of the process, but aim to calculate only the final result, i.e. the scattering matrix. The method of dispersion relations tries to do this.

The dispersion relation method does not attempt to represent by a wave function the state of a system at all times. Wave functions come into the theory only in order to describe the motion of the interacting particles before and after they scatter. Within the frame work of dispersion relations there is no such thing as the wave function describing the internal structure of a proton (or even a hydrogen atom), and the theory considers questions on the structure of an elementary particle to be meaningless. Only the scattering matrix is significant. Almost all the classical mechanical aspects have disappeared in the dispersion relation method. Only conservation laws, such as conservation of energy, momentum, angular momentum, have survived.

Dispersion relations have not answered all questions concerning the interactions of elementary particles, but they have achieved some important contributions. In elementary particle physics they were applied first for analysing the results of pion–nucleon scattering experiments. It was found that the results of the static model of Chew and Low could be derived from very general assumptions using dispersion relations. Several new ways of determining the renormalized pion–nucleon coupling constant were developed, but the values found were quite close to the ones quoted earlier in this section. Chapter 5 of Moravcsik's (1963) book contains an introduction to dispersion relations as applied to nucleon–nucleon scattering.

In this section we have used the "rationalized" pseudoscalar coupling constant f. Sometimes an "unrationalized" coupling constant f' is used which is related to f by

$$f^2/4\pi = f'^2$$

The difference between the two is analogous to the difference between rationalized and unrationalized units in classical electromagnetic theory. Another rationalized coupling constant g is often used. It is related to f by the equation

$$f^2 = g^2\left(\frac{m}{2M}\right)^2$$

where m/M is the ratio of the meson mass to the nucleon mass. The ratio $M/m = 6{\cdot}72$ and $g^2/4\pi = 14{\cdot}5$ if $f^2/4\pi = 0{\cdot}08$. Historically, the coupling constants f and g occurred in two forms of the pseudoscalar meson theory which differed in the way the meson field was coupled with the nucleons. The two couplings were called pseudovector and pseudoscalar respectively. At one time it was hoped to distinguish between them, but the predictions of the two theories differed only where calculations were unreliable. The dispersion relation approach makes no distinction between the two theories.

6.6 The One-Pion Exchange Interaction

If we assume that Yukawa's idea is correct, the interaction energy of two nucleons separated by a large enough distance should be dominated by the one-pion exchange contribution

$$V(r) = \tfrac{1}{3}\frac{f^2}{4\pi}mc^2\,\boldsymbol{\tau}_1.\boldsymbol{\tau}_2 \times$$
$$\times\left[\boldsymbol{\sigma}_1.\boldsymbol{\sigma}_2 + S_{12}\left(1+\frac{3}{\mu r}+\frac{3}{(\mu r)^2}\right)\right]\exp(-\mu r)/\mu r \qquad (6.23)$$

where f is the renormalized pion–nucleon coupling constant and μ is related to the meson mass m by the equation $\mu = mc/\hbar$. The numerical value of $\mu^{-1} = 1{\cdot}41$ fm. If the coupling constant f could be determined from experimental studies of the nucleon–nucleon interaction its value could be compared with the one obtained from meson scattering and photo production experiments. Such a comparison could verify Yukawa's theory quantitatively.

A systematic approach to the theoretical study of the meson exchange interaction was developed by a group of Japanese physicists in 1951 (Taketani, Nakamura and Sasaki). They recognized the importance of Wick's relation between the range of an exchange interaction and the total mass of the particles exchanged and realized that the two meson exchange contribution to the internucleon interaction energy should have a range half as big as that of the one meson exchange part. The three meson range should be one third as large and so on. Using this idea, they distinguished three regions in the internucleon exchange potential:

(i) An external region where the separation between the nucleons $r > 1{\cdot}5\mu^{-1} = 2{\cdot}1$ fm, in which one-pion exchange dominated. The interaction was correctly represented by the one-pion exchange potential (OPEP) (equation 6.23).
(ii) An intermediate region $2\ \text{fm} > r > 1\ \text{fm}$, where two-meson exchange was important. They estimated the potential in this region using fourth order perturbation theory.
(iii) An internal region where nothing could be calculated in a reliable way. The interaction here was represented by a phenomenological potential.

The first potential based on this approach was calculated by Taketami, Machida and Onuma (1952). In the first part of their paper they investigated a pure one-pion exchange interaction with a repulsive core. They varied the coupling constant f and the radius of the hard core until they got values for the deuteron binding energy and quadrupole moment which agreed with experimental ones. The coupling constant turned out to be

$$f^2/4\pi = 0{\cdot}075$$

which was very close to the value obtained from meson scattering experiments a few years later. The triplet effective range also agreed well with the experimental value. The same coupling constant did not give a good value for the singlet effective range, but Taketani and his collaborators knew that their potential was quite unrealistic in the two-pion exchange region, i.e. for nucleon separations less than 1·5 fm. They then calculated the two-pion exchange potential with fourth order perturbation theory. Using the one-pion and two-pion exchange potentials for large separations and a phenomenological interaction for the internal region of the potential they could fit the deuteron parameters and the singlet and triplet scattering parameters with a coupling constant

$$f^2/4\pi = 0{\cdot}09$$

Studies of the two-pion exchange interaction were continued by many other groups using different approximations. Some of these potentials have been summarized in graphical form by Moravcsik (1963). His graphs show clearly that the various theoretical potentials are different in the two-pion exchange region. By 1956 the Japanese workers in this field realized that calculations in the two-pion exchange region were quantitatively unreliable. The only unambiguous prediction seemed to be for the one-pion exchange region. Were there any properties of the two-nucleon system that were sensitive to the potential in this outer region?

In 1956 Iwadare *et al.* showed that the deuteron quadrupole moment was very sensitive to the internucleon potential in the one-pion exchange region, and much less sensitive to the potential in the two-pion or many-pion exchange regions. In order to give the correct values for the deuteron binding energy and quadrupole moment and the neutron–proton triplet effective range the pion–nucleon coupling constant must satisfy the condition

$$0{\cdot}065 < f^2/4\pi < 0{\cdot}09$$

independent of the interaction in regions (ii) and (iii). This restriction on the pion–nucleon coupling constant derived from properties of the deuteron was consistent with values obtained from meson scattering and meson photo-production. The value for the coupling constant derived by Taketani *et al.* using the one-pion exchange potential with a hard core was so good because the deuteron quadrupole moment was insensitive to details of the interaction in the inner regions.

More recently the one-pion exchange interaction has been used in the phase shift analysis of high energy nucleon–nucleon scattering experiments. As shown in Chap. V the scattering in states with large orbital angular momentum is only influenced by the long range part of the internucleon interaction. The phase shifts in these states should be determined by the one-pion exchange interaction. In a recent analysis by Breit (1960) the pion–nucleon coupling constant was taken as a variable parameter in the phase shift analysis and its value determined to give the best fit to the scattering data. An analysis of all the experimental data on proton–proton and neutron–

proton scattering from 217 to 350 MeV gave

$$g^2/4\pi = 14{\cdot}5 \pm 0{\cdot}3$$

or

$$f^2/4\pi = 0{\cdot}080 \pm 0{\cdot}002$$

We thus see that determinations of the pion–nucleon coupling constant from pion–nucleon scattering experiments, pion photo-production experiments, nucleon–nucleon scattering experiments and from properties of the deuteron are all consistent with each other. This fact is a striking verification of Yukawa's theory of nuclear forces.

6.7 Phenomenological Potentials

In the first few chapters of this book we showed that there has always been a tendency to describe the two-nucleon interaction by a potential. Modern theoretical approaches, however, based on dispersion relations do not use the concept of a potential, and it is certainly true that the potential energy of two nucleons cannot be measured directly, but only inferred by analysing scattering experiments and properties of the deuteron. Future theories of elementary particle interactions may show that the potential concept is not relevant, although at present the potential representation of the two-nucleon interaction is still useful for studies of nuclear structure.

There have been many attempts to find phenomenological potentials that fit the nucleon–nucleon scattering data and properties of the deuteron. Some of these have been based on meson theory, while some were purely empirical.

All the phenomenological potentials have a hard core and a tensor force component. Recent work shows that a spin orbit force is also necessary to obtain a good fit to the scattering data.

Recently, potentials have been described by Hamada and Johnston (1962) and Lassila *et al.* (1962). Both papers incorporate the one-pion exchange potential and give a good description of the nucleon–nucleon scattering data. The parameters of the Hamada–Johnston potential are tabulated by Wilson (1963) and by Moravcsik (1963).

6.8 The Role of Heavy Mesons

Soon after the discovery of the π-meson the K-meson was found in cosmic rays, but it could not contribute to nuclear forces in a simple way because of conservation of strangeness. If a nucleon emits a K-meson it changes into a Λ or Z hyperon, hence the simplest exchange process that can occur involves the exchange of two K-mesons and an example is illustrated in Fig. 14. The mass of the

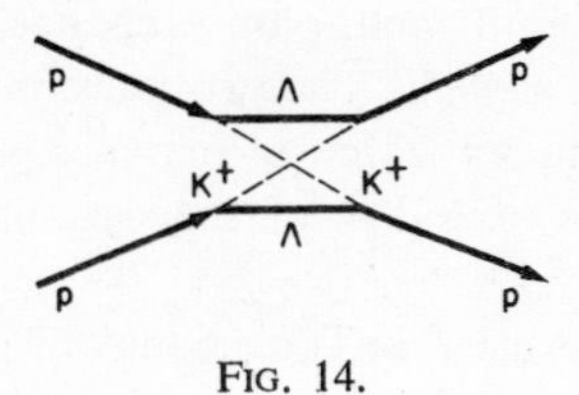

FIG. 14.

K-meson is 3·5 times that of the π-meson, so the range of the force due to exchange of a K-meson must be seven times shorter than the range of the one-pion exchange interaction.

Recently a number of other heavy mesons have been discovered among the products of reactions initiated by very high energy nucleons or pions. Some of them are listed in Table 6 with their quantum numbers (spin J, isobaric spin T and parity) and mass expressed in units of the pion mass. They all have a very short half life and are heavier than the pion. There is no reason, however, why

TABLE 6

Some well-established Mesons with Strangeness Zero

Meson	Spin	Isobaric Spin	Parity	Mass in units of pion mass
π	0	1	−	1
ABC	0	0	+	2·3
η	0	0	−	3·93
ρ	1	1	−	5·4
ω	1	0	−	5·6
ϕ	1	0	−	7·3

they should not contribute to the short range part of the nucleon–nucleon interaction. All have strangeness quantum number zero and can give rise to one-meson exchange interactions. It would seem to be inconsistent to include the effects of heavy meson exchange and neglect the two-pion or three-pion exchange interaction. There is some indication, however, that these heavy mesons might be composite particles made up of pions in some sort of way, and one might hope that one heavy meson exchange interactions already include the most important effects of multi-pion exchange. This is the basis of the *one boson exchange model.* The two-nucleon interaction potential is assumed to be a superposition of one-meson exchange potentials corresponding to the exchange of the various heavy mesons that have been observed.

The various mesons listed in Table 6 may be grouped according to Kemmer's classification (cf. Section 6). The ABC particle is a scalar meson, the η is a pseudoscalar meson and the ρ, ω and ϕ are vector mesons. The corresponding one-meson exchange potentials† were first derived by Kemmer in 1938 and are listed in Section 6.

Two striking qualitative features of the nucleon–nucleon interaction are the repulsive core and the strong spin-orbit coupling. These effects cannot be explained simply by π-meson exchange, but they are natural characteristics of a one vector-meson exchange interaction (Breit 1959). The ω and ϕ are both $T=0$ vector mesons, whereas the ρ is a $T=1$ vector meson. The vector meson exchange interaction depends on two parameters, the longitudinal and transverse coupling constants f_l and f_t. Provided $f_l^2 > 2f_t^2$ the central part of the $T=0$ exchange interaction is always repulsive (equation 6.16). Hence, the repulsive core in the internucleon interaction could be due to exchange of ω or ϕ mesons. Vector meson exchange also gives rise to a spin–orbit coupling (eq. 6.18).

† These potentials were calculated in the static limits assuming the meson to be much lighter than the nucleon. They should be corrected for the effects of nucleon recoil (Bryan 1963, Wong 1963).

Bibliography

ADAIR, R. K., *Phys. Rev.* **87,** 1041 (1952).
ALFORD, W. P. and FRENCH, J. B., *Phys. Rev. Letters,* **6,** 119 (1961).
ANDERSON, C. D., *Science,* **76,** 238 (1932), *Phys. Rev.,* **43,** 491 (1933).
ANDERSON, C. D. and NEDDERMEYER, S. H., *Phys. Rev.* **51,** 884 (1937).
ASTON, F. W., *Proc. Roy. Soc.* **A115,** 487 (1927).
AUFFRAY, G. P., *Phys. Rev. Letters,* **6,** 120 (1961).
BARTLETT, J. H., *Phys. Rev.* **49,** 102 (1936).
BAYER, HAHN and MEITNER, L., *Phys. Zeit.* **12,** 273, 378 (1911).
BERNARDINI, G. and GOLDWASSER, E. L., *Phys. Rev.* **54,** 436 (1955).
BETHE, H., *Phys. Rev.* **54,** 436 (1938).
BETHE, H., *Phys. Rev.* **55,** 1261 (1939).
BETHE, H., *Phys. Rev.* **76,** 38 (1949).
BETHE, H. and BACHER, R. F., *Rev. Mod. Phys.* **8,** 82 (1937).
BETHE, H. and MARSHAK, R. E., *Phys. Rev.* **72,** 506 (1947).
BHABHA, H. J., *Nature,* **143,** 276 (1939).
BIELER, E. S., *Proc. Roy. Soc.* **A105,** 434 (1924).
BJERGE, T. and WESTCOTT, C. H., *Proc. Roy. Soc.* **A150,** 709 (1935).
BLATT, J. M. and JACKSON, J. D., *Phys. Rev.* **76,** 18 (1950).
BLATT, J. M. and JACKSON, J. D., *Rev. Mod. Phys.* **22,** 77 (1950).
BOHR, N., *J. Chem Soc.* **349** (1932).
BREIT, G., *Phys. Rev.* **120,** 287 (1960).
BREIT, G., *Proc. nat. Acad. Sci. U.S.* **46,** 746 (1960).
BREIT, G., *Rev. Mod. Phys.* **34,** 766 (1962).
BREIT, G., CONDON, E. U. and PRESENT, R. D., *Phys. Rev.* **50,** 825 (1936).
BREIT, G., THAXTON, H. M. and EISENBUD, L., *Phys. Rev.* **55,** 1018 (1939).
BROWN, A. B., CHAO, C. Y., FOWLER, W. A., LAURITSEN, C. C., *Phys. Rev.* **78,** 88 (1949).
BRUECKNER, K. A. and WATSON, K. M., *Phys. Rev.* **92,** 1023 (1953).
BRYAN, R. A., DISMUKES, C. R. and RAMSAY, W., *Nuc. Phys.* **45,** 353 (1963).
CHADWICK, J., *Nature,* **129,** 312 (1932).
CHADWICK, J. and BIELER, E. S., *Phil. Mag.* **42,** 923 (1921).
CHEW, G. F. and LOW, F. E., *Phys. Rev.* **101,** 1570, 1579 (1956).
COCKCROFT, J. D. and WALTON, E. T. S., *Proc. Roy. Soc.* **A129,** 477 (1932).
COCKCROFT, J. D. and WALTON, E. T. S., *Proc. Roy. Soc.* **A137,** 229 (1933).
CONDON, E. U. and GURNEY, R. W., *Nature,* **122,** 439 (1928).
CONVERSI, M., PANCINI, E. and PICCIONI, O., *Phys. Rev.* **71,** 209 (1947).
COOPER, L. M. and HENLEY, E. M., *Phys. Rev.* **92,** 801 (1953).

DIRAC, P. A. M., *Proc. Roy. Soc.* **A114,** 243 (1927).
DUNNING, J. R., PEGRAM, G. B., FINK, G. A. and MITCHELL, D. P., *Phys. Rev.* **47,** 970 (1935).
ECKART, C., *Phys. Rev.* **35,** 1303 (1930).
EHRENFEST, P. and OPPENHEIMER, J. R., *Phys. Rev.* **37,** 333 (1931).
ELLIS, C. D. and WOOSTER, W. A., *Proc. Roy. Soc.* **A117,** 109 (1927).
FERMI, E., *Rev. Mod. Phys.* **4,** 84 (1932).
FERMI, E., Rapports du 7e Conseil de Physique, Solvay, p. 333 (1933).
FERMI, E., *Z. Phys.* **88,** 161 (1934).
FERMI, E., *Proc. Roy. Soc.* **A149,** 522 (1935).
FERMI, E., *Ric. Scientifica,* **7** (2), 13 (1936).
FERMI, E., TELLER, E. and WEISSKOPF, V., *Phys. Rev.* **71,** 314 (1947).
FERRELL, R. A. and VISSCHER, W. M., *Phys. Rev.* **107,** 781 (1957).
FEYNMAN, R. P., *Phys. Rev.* **76,** 749, 769 (1949).
FITCH, V. L. and RAINWATER, J., *Phys. Rev.* **92,** 789 (1953).
FOWLER, W. A., DELSASSO, L. A. and LAURITSEN, C. C., *Phys. Rev.* **49,** 561 (1936).
FRÖHLICH, H., HEITLER, W. and KEMMER, N., *Proc. Roy. Soc.* **A166,** 154 (1938).
GAMOW, G., *Z. Phys.* **51,** 204 (1928).
GAMOW, G., *Proc. Roy. Soc.* **A126,** 632 (1928a).
GARDNER, E. and LATTES, C. M. G., *Science,* **107,** 270 (1948).
HALPERN, J., *Phys. Rev.* **52,** 142 (1937).
HAMADA, T. and JOHNSTON, I. D., *Nuc. Phys.* **34,** 382 (1962).
HAMERMESH, M., *Phys. Rev.* **77,** 140 (1950).
HEISENBERG, W., *Z. Phys.* **77,** 1 (1932).
HEISENBERG, W., *Z. Phys.* **80,** 587 (1933).
HEISENBERG, W., Rapports du 7e Conseil de Physique, Solvay, p. 289 (1933a).
HEITLER, W. and HERTZBERG, G., *Naturwissenschaften,* **17,** 673 (1929).
HERB, R. G., KERST, D. W., PARKINSON, D. B. and PLAIN, A. J., *Phys. Rev.* **55,** 998 (1939).
HUGHES, D. J., BURGY, M. T. and RINGO, G. R., *Phys. Rev.* **77,** 291 (1950).
IWADARE, J., OTSUKI, S., TAMAGAKI, R. and WATARI, W., *Progr. theor. Phys.* **16,** 455 (1956).
IWANENKO, D., *Nature,* **129,** 798 (1932).
JASTROW, R., *Phys. Rev.* **81,** 165 (1951).
JENSEN, J. H. D., *Phys. Rev.* **75,** 1766 (1949).
KELLOGG, J. M. B., RABI, I. I., RAMSEY, N. F. and ZACHARIAS, J. R., *Phys. Rev.* **55,** 318 (1939).
KELLOGG, J. M. B. and RABBI, I. I., *Phys. Rev.* **57,** 677 (1940).
KEMMER, N., *Proc. Camb. phil Soc.* **34,** 354 (1938).
KEMMER, N., *Proc. Roy. Soc.* **A166,** 127 (1938a).
KONUMA, M., MIYAZAWA, H. and OTSUKI, S., *Progr. theor. Phys.* **19,** 17 (1957).
KROLL, N. M. and FOLDY, L. L., *Phys. Rev.* **88,** 1177 (1952).
KROLL, M. and RUDERMAN, M. A., *Phys. Rev.* **93,** 233 (1954).
KRONIG, R., *Naturwissenschaften,* **16,** 335 (1928).
KRONIG, R., Bandspectra & Molecular Spectra, Section 18, p. 94, C.U.P. (1930).
LASSILA, K. E., HULL, M. H., RUPPEL, H. M., MCDONALD, F. A. and BREIT, G., *Phys. Rev.* **126,** 881 (1962).

LATTES, C. M. G., MUIRHEAD, H., OCCHIALINI, G. P. S. and POWELL, C. F., *Nature*, **159,** 694 (1947).
LAURITSEN, T., FOWLER, W. A. and LAURITSEN, C. C., *Nucleonics*, **18** (1948).
LAWRENCE, E. O. and LIVINGSTON, M. S., *Phys. Rev.* **37,** 1707 (1931).
LAWRENCE, E. O. and LIVINGSTON, M. S., *Phys. Rev.* **40,** 19 (1932).
LAWRENCE, E. O. and LIVINGSTON, M. S., *Phys. Rev.* **42,** 150 (1932).
MAJORANA, E., *Z. Phys.* **82,** 137 (1933).
MAYER, M. G., *Phys. Rev.* **75,** 1969 (1949).
MCMILLAN, E. M., *Phys. Rev.* **68,** 143 (1945).
MESSIAH, A., *Quantum Mechanics*, North-Holland, Amsterdam (1962).
MORAVCSIK, M., *The Two-Nucleon Interaction*, O.U.P. (1963).
NORDHEIM, L. W. and YOST, F. L., *Phys. Rev.* **51,** 942 (1937).
OCCHIALINI, G. P. S. and POWELL, C. F., *Nature*, **159,** 186 (1947).
PANOFSKY, W. K. H., *Phys. Rev.* **81,** 565 (1951).
PAULI, W. E., *Viert. Naturf. Ges. Zürich*, **102,** 387 (1957), *Collected Scientific Papers*, Vol. II, 1313, *Interscience* (1964).
PERKINS, D. H., *Nature*, **159,** 126 (1947).
PRESTON, M. A. and SHAPIRO, J., *Canad. J. Phys.* **34,** 451 (1956).
RADICATI, L. A., *Phys. Rev.* **87,** 525 (1952).
RASSETTI, F., *Z. Phys.* **61,** 598 (1930).
RASSETTI, F., Collected Papers of E. Fermi, Vol. I, p. 538, U.C.P. (1962).
ROSENFELD, L., *Nuclear Forces*. New York: Interscience (1948).
RUTHERFORD, E., *Phil. Mag.* **21,** 669 (1911).
RUTHERFORD, E., *Phil. Mag.* **37,** 537 (1919).
RUTHERFORD, E., *Proc. Roy. Soc.* **A97,** 374 (1920).
RUTHERFORD, E., *Phil. Mag.* **4,** 580 (1927).
RUTHERFORD, E., *Proc. Roy. Soc.* **A136,** 735 (1932).
SAKATA, S. and INOUE, T., *Progr. theor. Phys.* **1,** 143 (1946).
SCHWINGER, J. S., *Phys. Rev.* **52,** 1250 (1937).
SCHWINGER, J. S., *Phys. Rev.* **72,** 742 (1947).
SCHWINGER, J. S. and TELLER, E., *Phys. Rev.* **52,** 286 (1937).
SCOTTI, A. and WONG, D. Y., *Phys. Rev. Letters*, **10,** 142 (1963).
SHERR, R., MUETHER, H. R. and WHITE, M. G., *Phys. Rev.* **75,** 282 (1948).
SOMMERFELD, A., *Atombau und Spectrallinien*, 538 Vieweg (1919).
SQUIRES, G. L. and STEWART, A. T., *Proc. Roy. Soc.* **A230,** 19 (1955).
TAKETANI, M., NAKAMURA, S. and SASAKI, M., *Progr. theor. Phys.* **6,** 581 (1951).
TAKETANI, M., MACHIDA, S. and ONUMA, S., *Progr. theor. Phys.* **7,** 45 (1952).
TAMM, J. and IWANENKO, D., *Nature, London*, **133,** 981 (1934).
THOMAS, R. G., *Phys. Rev.* **80,** 136 (1950).
TOMONAGA, S. and ARAKI, G., *Phys. Rev.* **58,** 90 (1940).
TRAINOR, L., *Phys. Rev.* **85,** 962 (1952).
TUVE, M. A., HAFSTAD, L. R. and DAHL, O., *Phys. Rev.* **43,** 1055 (1933).
TUVE, M. A., HEYDENBERG, N. and HAFSTAD, L. R., *Phys. Rev.* **50,** 806 (1936).
VAN PATTER, D. M., *Phys. Rev.* **76,** 1264 (1949).
VEKSLER, V. I., *J. Phys. U.S.S.R.* **9,** 153 (1945).
WHEELER, J. A., *Phys. Rev.* **71,** 320 (1947).
WHITE, M. G., *Phys. Rev.* **49,** 309 (1936).
WICK, G. C., *Nature*, **142,** 994 (1938).

WIGNER, E., *Phys. Rev.* **43,** 252 (1932).
WIGNER, E., *Z. Phys.* **83,** 253 (1933).
WIGNER, E., *Phys. Rev.* **51,** 106 (1937).
WIGNER, E., *Phys. Rev.* **56,** 519 (1939).
WIGNER, E. and EISENBUD, L., *Proc. nat. Acad. Sci. U.S.* **27,** 281 (1941).
WILKINSON, D. (1958).
WILSON, R., *The Nucleon–Nucleon Interaction* (1963).
WU, C. S., *Theoretical Physics in the 20th Century* (1960).
YUKAWA, H., *Proc. Phys.-Math. Soc. Japan*, **17,** 48 (1935).
YUKAWA, H., *Proc. Phys.-Math. Soc. Japan*, **19,** 1082 (1937).
YUKAWA, H., SAKATA, S. and TAKETANI, M., *Proc. Phys.-Math. Soc. Japan*, **20,** 319, 720 (1938).

Part 2

1

Discussion on the Structure of Atomic Nuclei†

OPENING ADDRESS

By Lord RUTHERFORD, O.M., F.R.S.

LORD RUTHERFORD: In my address today, I shall briefly review some of the main lines of advance in our knowledge of atomic nuclei since the last discussion‡ which I had the honour to open. In the interval, there has been substantial progress in many directions, and new and promising methods of attack on this formidable problem have been opened up. I can only refer in passing to the valuable data obtained by Aston and others on the isotopic constitution of the elements and the relative abundance of the isotopes of many of the elements. This has made it possible to determine the chemical atomic weight of many elements with considerable accuracy by the use of the mass spectrograph. A number of new experiments have been made to determine with accuracy the relative quantities of the isotopes of lead, and in particular of lead obtained from pure uranium and thorium minerals of great geological age. Data of this kind are of much interest and importance, not only from the point of view of radioactivity but also with regard to the fixation of an accurate time scale in geology. It seems certain that the end product of the actinium series—actinium lead—has an atomic mass 207 and that actinium is derived from the transformation of an isotope of uranium. From the relative abundance of actinium-lead and uranium-lead derived

† *Proc. Roy. Soc.*, **A136**, 735 (1932).
‡ *Proc. Roy. Soc.*, **A123**, 373 (1929).

from old radioactive minerals, it is possible to deduce the average life of this uranium isotope. I pointed out some time ago in *Nature*, that important inferences could be drawn with regard to the production of elements in the sun from a consideration of the average life of the two uranium isotopes.

Optical Methods

One of the most interesting developments in recent years has been the application of optical methods to determine the presence of isotopes and to throw light on the movements of the nucleus. The study of the band-spectra of the molecules of the lighter elements had disclosed the presence of isotopes existing in small quantity compared with the main isotope. It has been shown that oxygen consists of three isotopes of masses 16, 17, 18, carbon 12, 13, beryllium 8, 9, boron 11, 10, while recent observations of Urey, Brickwedde and Murphy are believed to indicate the presence in small quantity in hydrogen of a new isotope of mass 2. Attempts are in progress to concentrate the new isotope by fractional distillation of liquid hydrogen.

In addition to identifying lines due to new isotopes, considerable attention has been directed to the relative intensities of the lines in the band spectra. Not only does this yield information about the spin of the nucleus but it also provides a method of attack on one of the most fundamental points of nuclear physics, namely whether the members of a given isotopic system are identical.

During the last few years, there have been many researches to determine the hyperfine structure in optical spectra. This gives another line of attack on the difficult problem of spin of the nucleus. I shall leave to Professor R. H. Fowler the task of discussing the data obtained and the conclusions that can be drawn.

Application of Wave Mechanics

In the last discussion reference was made to the application by Gamow and by Gurney and Condon of the then new wave-mechanical ideas to certain nuclear problems, and in particular to the explanation of the well-known Geiger–Nuttall rule connecting the velocity of escape of an α-particle from a radioactive substance and its transformation constant. On this theory it was supposed that the nucleus was surrounded by a high positive potential barrier, and that the α-particles and other nuclear constituents within this barrier were held in equilibrium by strong but unknown types of attractive forces. On such a model, there is a finite probability that the α-particle in the nucleus can escape through the barrier without loss of energy, the probability increasing rapidly with increase of the energy of the α-particle. This general conception of the nucleus has proved very valuable in a number of directions and has been a very useful working guide to the experimenter. Unfortunately it has not so far been found possible on the theory to give any detailed picture of the structure of a nucleus. It is generally supposed that the nucleus of a heavy element consists mainly of α-particles with an admixture of a few free protons and electrons, but the exact division between these constituents is unknown. On the theory, there is a great difficulty in including within the minute nucleus particles of such widely different masses as the α-particles and electron. In addition, the nucleus is such a concentrated structure, and the constituent particles are so close together, that the theory of the action of one particle on another, applicable under ordinary conditions, cannot be safely applied for such minute distances.

It appears as if the electron within the nucleus behaves quite differently from the electron in the outer atom. This difficulty may be of our own creation for it seems to me more likely that an electron cannot exist in the free state in a stable nucleus, but must always be associated with a proton or other possible massive units. The indication of the existence of the neutron in certain nuclei is significant in this connection. The observation of Beck, that in the building up of heavier elements from the lighter, electrons are added in pairs, is of much interest and suggests that it is essential to neutralize the large magnetic moment of the electron by the addition of another for the formation of a stable nucleus. It may be that uncharged units of mass 2 as well as the neutron of mass 1 may be secondary units in the structure of nuclei.

While no definite theory of the nucleus seems at present feasible, yet much progress can be made by the use of suitable analogies based on the general model of the nucleus previously outlined. For example, Gamow has drawn important inferences with regard to the mass defect of the lighter atoms composed of α-particles, i.e. of elements of the type $4n$, on the analogy that the forces in the nucleus resemble in a general way those acting in a minute drop of water. In addition he has discussed in an illuminating way the conditions to be fulfilled for the formation of stable nuclei of high atomic number. Unfortunately, the masses of the isotopes of many of the elements require to be known with much greater accuracy before much further progress can be made in this problem along these lines.

In another direction, too, it has been found fruitful to apply to the nucleus many of the general ideas of energy levels which have proved so useful in discussing the electronic structure of the outer atom. It has long been supposed that the quantum laws hold within the nucleus, and the correctness of this assumption has been abundantly verified in recent years. It will be seen that the conceptions of energy levels and of excitation of nuclei have proved of great utility in much recent work on the difficult problem of the origin of the γ-rays, and in understanding the observations obtained in the study of the artificial disintegration of the elements.

Origin of the γ-Rays

It has long been recognized that the γ-rays originate in the nucleus and represent in a sense the characteristic modes of vibration of the nuclear structure. The interpretation of the complicated γ-ray spectra shown by the radioactive elements has, however, been rendered difficult by our ignorance of the origin of this radiation—whether it arises from the constituent electron, proton or α-particle, or from the nucleus acting as a single entity. During the last few years, there has been a vigorous attack on this problem, and it now seems clear that the nuclear γ-rays are due to the transition of an α-particle between energy levels in an excited nucleus. Two different lines of attack have been developed depending on:

(1) A study of the long range α-particles from radium C and thorium C;

(2) The fine structure shown in the α-particles emission from certain radioactive substances.

It may be supposed that the emission of a β-particle during a transformation causes a violet disturbance in the resulting nucleus, some of the constituent α-particles being raised to a much higher level of energy than the normal. These α-particles are unstable, and after a very short interval fall back to the normal level, emitting their surplus energy in the form of a γ-ray of definite frequency, defined by the quantum relation. In this short interval, there is a small chance that some of the α-particles in the higher levels may escape through the potential barrier of the nucleus. On this point of view, the escaping α-particles from the different levels represent the groups of long range α-particles observed. The energy of the escaping α-particle gives the value of the energy level occupied by the α-particle before its release in the excited nucleus.

To test this hypothesis, the long range α-particles from radium C have been carefully analysed, using the new counting methods, by a group of workers, Wynn Williams, Ward, Lewis and the writer, and found to consist of at least nine distinct groups. The difference of energy between the various groups was found to be closely connected with the energies of some of the most prominent γ-rays, and in general the experiments gave strong evidence that the γ-rays had their origin in the transitions of one or more α-particles in an excited nucleus. At the same time, the experiments gave us direct information of the magnitude of a number of the possible energy levels in this particular nucleus.

In the great majority of cases, the α-particles in a radioactive transformation are expelled with identical speed. Rosenblum, however, showed that the element thorium C emitted not one but five distinct groups of α-particles, and evidence of a fine structure in the α-rays has since been obtained for other radioactive bodies. Gamow pointed out that γ-rays should arise in all cases where such a fine structure in the α-rays was present. Owing to certain technical difficulties in the case of thorium C, it has been difficult to give clear-cut evidence of the correctness of this point of view. Ellis and also Rosenblum conclude that Gamow's view is correct, but Meitner reached an opposite conclusion.

I can only refer in passing to some recent experiments of Mr. Bowden and myself for the proof of the emission of γ-rays from the actinium emanation, which Lewis and Wynn Williams found emitted two distinct groups of γ-rays. The results seem to me to support the general correctness of the theory that a fine structure in the emission of α-rays is always accompanied by the appearance of γ-rays. I shall leave to a later speaker, Dr. Ellis, the task of dealing more adequately with the present situation of this important problem.

When once the origin of the γ-rays is definitely settled there is a reasonable prospect of attacking successfully the whole question of the interpretation of γ-ray spectra in general, in which so far only a beginning has been made. It is obvious that a closer understanding of this problem may be expected to throw much light on the detailed structure of the nucleus. It is of great importance for this purpose to determine the spectrum of the γ-rays with the greatest possible precision, and this will entail many years of work.

Before leaving this part of the subject, I should like to emphasize the remarkable difference between the emission of an α-particle and a β-particle in disturbing a nucleus. Strange to say, the escape of an α-particle either does not excite a nucleus at all, or only raises one more of the constituent α-particles to a comparatively low level of energy above the normal. In many cases, however, the escape of a β-particle causes a violent excitation of the residual nucleus, some of the α-particles being raised to a very high level of energy and with the emission of high energy γ-rays. This difference between the effects of the two types of particle is very striking and may be intimately connected with the processes which cause the emission of a β-particle from a radioactive element.

Whenever we have to deal with the behaviour of the electron in the nucleus, we find grave difficulty in applying our theoretical ideas. The most striking instance is perhaps that radioactive nuclei of the β-ray type emit electrons with a wide range of energy, and that there appears to be no compensating process which would permit of that definite energy balance to be expected on quantum dynamics. This is undoubtedly one of the most fundamental problems today, but it is unlikely that we shall have sufficient time to discuss its theoretical implications.

Excitation of Nuclei by γ-Rays

Until recently, it had been generally supposed that the absorption of X-rays and γ-rays was due entirely to the interaction of the radiation with the extra nuclear electrons, and that the nucleus itself took no part in the process. It is now clear that if the quantum energy of the γ-rays exceeds about 2 MeV, an additional type of absorption appears with ordinary nuclei, accompanied by the emission of characteristic radiations of different frequencies from the primary. This effect of the nucleus on the absorption has been brought out by the work of Chao, Meitner, Hupfield, Tarrant and others using the penetrating γ-radiation from thorium C of energy about $2{\cdot}65 \times 10^6$ eV. In a paper now in course of publication by this Society, Gray and Tarrant† give the results of a detailed examination of this nuclear excitation by different elements. Not only γ-rays from thorium C have been used, but also the high frequency components in the radiation from radium C. They conclude that this nuclear excitation is a general property of the elements at any rate between oxygen and lead. Characteristic radiations of similar type appear to be emitted by all the elements, the intensity of the radiation from different elements varying approximately as the square of the atomic number. These characteristic radiations from the nucleus which appear to be emitted uniformly in all directions, can be resolved into two components of quantum energy about 500,000 and 1,000,000 eV. In explanation they suggest that the γ-radiation does not excite the nucleus as a whole, but some constituent like the α-particle which is common to all the elements. It may be that the characteristic radiations observed represent some of the modes of vibration of the α-particle structure itself. It will be of much interest to pursue these important investigations further, but progress is hindered by the difficulty of obtaining

† *Proc. Roy. Soc.*, **A186,** 662.

strong sources of high frequency radiation over a wide range of quantum energy. The excitation of the nucleus by high frequency radiation is no doubt intimately connected with the processes which give rise to the γ-rays from a radioactive nucleus, and may help to throw further light on this problem.

Artificial Transmutation

In the last few years there has been a rapid increase of our knowledge of the artificial transmutation of light elements by bombardment with α-particles. This has been largely due to the development of new electrical methods of counting α-particles and protons in place of the useful but trying scintillation method. Pose first showed that some of the protons ejected from aluminium appeared in groups of definite velocities. Our knowledge has been extended by the work of Pose, Meitner, Bothe, de Broglie and Ringuet and Chadwick and Constable. For example, Chadwick and Constable have resolved the protons liberated from aluminium by the α-particles of polonium into eight distinct groups connected in pairs. In explanation it is supposed that the protons or α-particles in the bombarded nucleus occupy definite energy levels. It is supposed, following the suggestion of Gurney, that there is a much greater chance of entering the potential barrier of the nucleus owing to resonance if the bombarding α-particle has about the same energy as that of the proton or α-particle in the nuclear level. For a given energy of α-particle, two groups of protons are emitted corresponding, it is believed, to two distinct processes of capture of the α-particle by the nucleus. Similar results have been observed in fluorine and other light elements. It has been found that these resonance levels for privileged capture of the α-particle are fairly broad, corresponding to about 5 per cent of the energy level. The results as far as they have gone have yielded important information on the values of the energy levels of light nuclei, and with the use of still swifter particles than those from polonium we may expect to extend our knowledge of these levels still further.

In interpretation of these experiments, it has been implicitly assumed that the laws of the conservation of energy and momentum apply. In this way it has been possible to calculate with considerable accuracy the atomic mass of the element resulting from the capture of the α-particle and the emission of the proton.

When two groups of protons of different speeds are connected with a single resonance level, it is found that γ-rays appear of quantum energy corresponding approximately to the difference of energy of the protons in the two groups. The study of the γ-radiations emitted during artificial transmutation has in the last few months led to new and interesting developments. Bothe and Becker in 1930 found that the element beryllium when bombarded by α-particles did not emit protons but gave rise to what appeared to be a γ-radiation of penetrating power greater than the γ-rays from radium C′.

The absorption of this radiation by matter was examined by Mme. Curie-Joliot and M. Joliot, and also by Webster. This year Mme. Curie-Joliot and M. Joliot observed by the ionisation method that this radiation emitted protons of high velocity from hydrogen material. It was at first suggested that the swift protons might be due to an interaction between the γ-ray quantum and the proton, but this required the quantum energy of the radiation to be very high, of the order of 50 MeV. As a result of further experiments by electrical methods of counting, Chadwick found that a similar recoil effect could be observed for all light atoms, and concluded that the effects might be explained on the assumption that a stream of swift neutrons were liberated from the beryllium nucleus. It is not easy to distinguish between these two suggestions, but sufficient evidence has been accumulated to show that this new type of radiation has surprising properties and is able to produce disintegration in nitrogen, probably in a novel way.

I shall leave to Dr. Chadwick to give you a fuller account of the work on artificial transmutation and on the properties of this new type of radiation.

The idea of the possible existence of "neutrons," that is, of a close combination of a proton and an electron to form a unit of mass nearly one and zero charge is not new. In the Bakerian Lecture before this Society in 1920, I discussed the probable properties of the neutron, while the late Dr. Glasson and J. K. Roberts in the Cavendish Laboratory made experiments to detect the formation of neutrons in a strong electric discharge through hydrogen but without success. If the neutron hypothesis be confirmed by experiment, it will obviously much influence our conception of the formation and constitution of nuclei. Many years ago in a lecture before the Royal Institution, I discussed the possibility of the formation of heavy nuclei from hydrogen through the intermediary of the neutron. It seems not unlikely that the neutrons, owing to their mutual attraction, may collect in massive aggregates which in course of time by the processes of disintegration and association, re-arrange themselves to form the nuclei of the stable elements. I merely throw out this old idea as one possibly worthy of further consideration in the light of later knowledge.

Scattering of α-Particles

In previous discussions attention has been directed to the anomalous scattering of α-particles by light elements, and the difficulty of interpretation of the results obtained. Many of these difficulties have been removed by the application of the wave-mechanical ideas to these problems. For example, H. M. Taylor has been able to account in considerable detail for the anomalous scattering of α-particles observed both in hydrogen and helium, by simple considerations based on the wave-mechanics. Mott drew attention to the anomalies to be expected in the scattering of low velocity α-particles by helium, and his conclusions have been amply confirmed by the work of Chadwick and Blackett and Champion. On the theory of Mott, similar anomalies are to be expected in collisions between two identical nuclei of any kind.

General

I have endeavoured in the above review to bring to your attention what appear to me the most important lines of experimental attack on the problem of the structure of atomic nuclei. I have not entered into speculative questions like the possibility of the annihilation of matter and its conversion into radiation, nor have I referred to the guesses, which we may hope will prove inspired, of the numerical relation between the unit of charge and Planck's constant h or of the relation between the masses of the electron and proton; nor have I entered, except in an incidental way, upon the difficult question on which much has been written of the formation and transformation of nuclei, under the influence of conditions existing in hot stars.

In making this review, I have been struck by the comparatively rapid progress that has been made since our last discussion on the attack on this central problem of Physics. Progress would be much hastened if we could obtain in the laboratory powerful but controllable sources of high-speed atoms and high-frequency radiations to bombard matter. By the experiments of Tuve, Hafsted and Dahl, in the Department of Terrestrial Magnetism, Washington, and by Cockcroft and Walton in the Cavendish Laboratory, it has been found possible by the use of high potentials to produce a stream of protons artificially with an individual energy energy of about 1 MeV and to examine their properties. A number of other methods of producing high-speed atoms are under trial by other investigators, and I would especially refer to the ingenious method developed by Lawrence and Livingston, of the University of California, where, by using multiple accelerations, protons have been obtained of an energy corresponding to about 1 MeV. In a recent paper they conclude that it should be possible by this method to obtain a stream of high-speed atoms of a much higher energy. There is thus a hopeful prospect that we may be able in the near future to obtain useful sources of high-speed atoms and high frequency radiation, and thereby extend our knowledge of the structure of the nucleus.

Addendum

Since this statement has been circulated to the Fellows of the Society, some interesting new experiments have been made by J. D. Cockcroft and E. T. S. Walton in the Cavendish Laboratory. An apparatus has been designed to give a steady potential of 600,000–800,000 V. By means of an auxiliary discharge tube, protons are produced and then accelerated by a high potential in a vacuum tube.

In this way a steady stream of swift protons of energies up to 600,000 V can be produced and used to bombard a number of elements. The material to be bombarded by these swift ions was placed inside the tube at 45° to the direction of the beam. A thin mica window was sealed on to the side of the tube, and the existence of swift particles was investigated by the scintillation method outside the tube.

The first element examined was lithium, when a few bright scintillations were observed at an accelerating potential of about 125,000 V. The number increased rapidly with increase of voltage up to 400,000 V, when many hundreds of scintillations per minute were observed for a proton current of a few micro-amperes. These particles had a maximum range of about 8 cm in air. The brightness of the particles indicated that they were probably α-particles, and this was confirmed by observations of the tracks produced by these particles in an expansion chamber. It seems clear that some of the lithium nuclei have been disintegrated. The simplest assumption to make is that the lithium nucleus of mass seven captures a proton, and the resulting nucleus of mass eight breaks up into two α-particles. On this view the energy emitted corresponds to about 16 MeV, a value in good accord with the conservation of energy when we take into account the difference between the initial and final masses of the nuclei. If this view proves correct, a disintegrating nucleus of lithium should give rise to α-particles projected in opposite directions, and it is proposed to try experiments to test this. For about 200,000 V it can be estimated that the number of disintegrations is about 1 for 10^9 protons.

Experiments have been made with a number of other elements. Boron, fluorine and aluminium all give rise to particles, resembling α-particles, of a characteristic range for each element. A number of scintillations, some bright and others faint, were also observed from beryllium and carbon, and there is also an indication that nitrogen gives a few very bright scintillations. Oxygen and copper gave no scintillations for protons of energy up to 400,000 V.

It is obvious that a great amount of work will be required to examine all the elements by this method and to determine the nature of the swift particles which may be emitted. In some cases they appear to be α-particles, but we must always bear in mind the possibility of emission of particles of different types and masses.

It is not difficult to make suggestions as to the possible modes of disintegration of some of the elements mentioned consistent with the conservation of energy. For example, it may be possible that the nucleus of fluorine of mass 19, after capturing a proton, breaks up into an α-particle and the oxygen nucleus. Similarly, aluminium may be transformed into magnesium. We must, however, await further evidence before any definite decision can be reached on such questions. It is clear that the successful application of these new methods opens up a new and wide field of research where the effect of bombarding matter by swift ions of different kinds can be examined. Dr. Cockcroft and Dr. Walton are to be congratulated on their success in these new experiments which have taken several years of hard work in preparation.

J. Chadwick, F.R.S.: Experiments in which elements are bombarded by α-particles have proved particularly fruitful in providing information about the structure of nuclei. The advances since the last discussion have been partly due to improved technique in the experiments and partly to the application of the new mechanics to these problems. To show how this increased knowledge has been obtained I shall take as an example the case of the aluminium nucleus.

When a beam of α-particles falls on a thin foil of aluminium some of the particles are scattered by collisions with the aluminium nuclei. If the incident α-particles are slow, the scattering is completely described by the Rutherford theory of scattering, and we conclude that the force between the α-particle and the nucleus is given by Coulomb's law. As the velocity of the incident particles is increased, the scattering begins to depart from the normal laws; the amount of the scattering at 135°, for example, first decreases below the normal and then rapidly increases as the velocity of the α-particle is further increased. This anomalous scattering is difficult to explain by means of classical mechanics, but is easily accounted for on the wave mechanics. Suppose a particle makes a very close collision with the nucleus and comes to a point on the potential barrier where the thickness of the barrier is of the same order of magnitude as the wave-length of the α-particle. There is then a certain probability that the α-particle will penetrate the barrier. The scattered wave which represents such a particle will have a certain deviation in phase and will disturb the classical distribution of the scattered particles. The experiments of Riezler show that the scattering becomes anomalous when the α-particles come within a distance of 6×10^{-13} cm of the Al nucleus. On certain plausible assumptions it follows that the radius of the top of the potential barrier must lie between 3 and 6×10^{-13} cm. Taking a mean of $4{\cdot}5 \times 10^{-13}$ cm the height of the barrier of Al (against an α-particle) is about 8×10^6 eV.

I turn now to observations of the artificial disintegration of aluminium. When aluminium is bombarded by α-particles we observe, in addition to the scattered α-particles, an emission of protons of high energies which is roughly equal in all directions. An α-particle which penetrates into the nucleus of Al^{27} may be captured; a proton is emitted and a nucleus Si^{30} is formed. We assume that the α-particles and the protons in a nucleus are in definite energy levels. The captured α-particle of kinetic energy W, falls into a level E_a, say, and a proton is emitted from a level E_p (both below ground level). The kinetic energy of the ejected proton will be $W + E_a - E_p$, neglecting the small energy of the residual nucleus. On this view, a homogeneous beam of α-particles incident on a very thin Al foil should give rise to the emission of protons of the same energy (in a given direction). The observations show, however, that in such a case two groups of protons are emitted. This is explained by supposing that in some cases (the majority) the final Si^{30} nucleus is formed in two steps; the α-particle is captured (perhaps into an intermediate level) and a proton emitted with the formation of an excited Si^{30} nucleus, which passes into the ground state with the emission of a quantum of radiation. This explanation is supported by the observation that Al bombarded by α-particles does actually emit a γ-radiation of about the appropriate energy.

Observations of the protons emitted from a thick foil of Al exposed to polonium α-particles show that the protons consist of eight groups associated in pairs. Although collisions are taking place between Al nuclei and α-particles of all velocities from zero up to the initial velocity of the polonium α-particle, yet the disintegrations appear to be due only to α-particles of certain specified velocities. Such a possibility was first pointed out by Gurney, who suggested that there may be a resonance effect between the incident α-particle and the atomic nucleus. If the α-particle has exactly the energy corresponding to a resonance level of the nucleus its chance of penetrating the potential barrier will be very much greater than if its energy is more or less than this. The first evidence for this resonance effect was found by Pose in the disintegration of aluminium. The later observations just mentioned show that there are four resonance levels of the aluminium nucleus between about 4 and $5{\cdot}3 \times 10^6$ eV. The levels are not very sharp but have a width of about 250,000 V. Penetration of the α-particle through each level and its capture gives rise to the emission of a pair of proton groups.

There is still a large region of the potential barrier of aluminium which has not been investigated in this way. It may be that further experiment will discover certain relations between the levels of the same element and correspondences between the levels of one element and those of another.

Particular interest has recently been shown in the disintegration of the elements beryllium and boron. It was found by Bothe and Becker that these elements emitted a penetrating radiation, apparently of the γ type, when bombarded by polonium α-particles. A few months ago, Mme. Curie-Joliot and M. Joliot made the very striking observation that these radiations have the property of ejecting protons with high speeds from matter containing hydrogen. They found that the protons ejected by the beryllium radiation had velocities up to nearly 3×10^9 cm/sec. They supposed that the ejection of the protons occurred by a process analogous to the Compton effect, and concluded that the beryllium radiation had a quantum energy of about 50×10^6 eV. Two serious difficulties arise if this explanation be adopted. Firstly, it is known that the scattering of a quantum by an electron is well described by the Klein–Nishina formula, and there is no reason to suppose that a similar relation should not be true for scattering by protons. The observed scattering is, however, very much too great. Secondly, it is difficult to account for the emission of a quantum of such high energy from the transformation $Be^9 + He^4 \rightarrow C^{13} +$ quantum. I therefore examined the properties of this radiation, using the valve counter. It was found that the radiation ejects particles not only from hydrogen but from helium, lithium, beryllium, etc., and presumably from all elements. In each case the particles appear to be recoil atoms of the elements. It seemed impossible to ascribe the ejection of these particles to a recoil from a quantum of radiation, if energy and momentum are to be conserved in the collisions.

A satisfactory explanation of the experimental results was obtained by supposing that the radiation consists not of quanta but of particles of mass 1 and charge 0, or neutrons.

In the case of two elements, hydrogen and nitrogen, the ranges of the recoil atoms have been measured with fair accuracy, and from these their maximum velocities were deduced. They are $3{\cdot}3 \times 10^9$ cm/sec and $4{\cdot}7 \times 10^8$ cm/sec, respectively. Let M, V be the mass and maximum velocity of the particles of which the radiation consists. Then the maximum velocity which can be given to a hydrogen nucleus in a collision is

$$u_H = \frac{2M}{M+1} V$$

and to a nitrogen nucleus

$$u_N = \frac{2M}{M+14} V$$

Hence

$$\frac{M+14}{M+1} = \frac{u_H}{u_N} = \frac{3{\cdot}3 \times 10^9}{4{\cdot}7 \times 10^8}$$

and

$$M = 1{\cdot}15$$

Within the error of experiment M may be taken as 1, and therefore

$$V = 3{\cdot}3 \times 10^9 \text{ cm/sec}$$

Since the radiation is extremely penetrating the particle must have a charge very small compared with that of an electron. It is assumed that the charge is 0, and we may suppose that the neutron consists of a proton and an electron in close combination.

The available evidence strongly supports the neutron hypothesis. In the case of beryllium, the transformation process which results in the emission of a neutron is $Be^9 + He^4 \rightarrow C^{12} + n$. It can be shown that the observations are compatible with the energy relations of this process. In the case of boron, the transformation is probably $B^{11} + He^4 \rightarrow N^{14} + n'$. In this case, the masses of B^{11}, He^4, and N^{14} are known from Aston's measurements, the kinetic energies of the particles can be found by experiment, and it is therefore possible to obtain a much closer estimate of the mass of the neutron. The mass so deduced is 1·0067. Taking the errors of the mass measurements into account, it appears that the mass of the neutron probably lies between 1·005 and 1·008. Such a value supports the view that the neutron is a combination of proton and electron, and gives for the binding energy of the particles about 1 to 2×10^6 eV.

The neutron may be pictured as a small dipole, or perhaps better as a proton embedded in an electron. On either view the "radius" of the neutron will be between 10^{-13} cm and 10^{-12} cm. The field of the neutron must be very small except at close distances, and the neutrons in their passage through matter will be unaffected except when they make a direct hit on an atomic nucleus. Measurements made on the passage of neutrons through matter give results in general agreement with these views. The collisions of neutrons with nitrogen nuclei have been studied by Dr. Feather, using an automatic expansion chamber. He has found that, in addition to the normal tracks of nitrogen recoil atoms, there are a number of branched tracks. These are due to disintegration of the nitrogen nucleus. In some cases the neutron is captured and an α-particle is emitted, a nucleus of B^{11} being formed. In other cases the mechanism is not yet known with certainty.

2

Chemistry and the Quantum Theory of Atomic Constitution†

NIELS BOHR

THIS situation must above all be kept in mind when we turn to the problem of the *constitution of atomic nuclei*. The empirical evidence regarding the charges and the masses of these nuclei, as well as the evidence concerning the spontaneous and the excited nuclear disintegrations, leads, as we have seen, to the assumption that all nuclei are built up of protons and electrons. Still, as soon as we inquire more closely into the constitution of even the simplest nuclei, the present formulation of quantum mechanics fails essentially. For instance, it is quite unable to explain why four protons and two electrons hold together to form a stable helium nucleus. Evidently we are here entirely beyond the scope of any formalism based on the assumption of point electrons, as it also appears from the fact that the size of the helium nucleus, as deduced from the scattering of α-rays in helium, is of the same order of magnitude as the classical electron diameter. Just this circumstance suggests that the stability of the helium nucleus is inseparably connected with the limitation imposed on classical electrodynamics by the existence and the stability of the electron itself. This means, however, that no direct attack on this problem, based on the usual correspondence argument, is possible as far as the behaviour of the intra-nuclear electrons is concerned. As regards the behaviour of the protons, the situation is essentially different, since their comparatively large mass permits of an unambiguous use of the idea of space coordination even within nuclear dimensions. Of course, in absence of a general consistent

† *J. Chem. Soc.* 349 (1932).

theory accounting for the stability of the electron, we cannot make any direct estimate of the forces which hold the protons in the helium nucleus, but it is interesting to note that the energy liberated by the formation of the nucleus, as calculated from the so-called mass-defect by means of Einstein's relation, is in approximate agreement with the binding energy of the protons to be expected on quantum mechanics from the known nuclear dimensions. Indeed, this agreement indicates that the value of the ratio of the masses of the electron and the proton plays a fundamental part in the question of the stability of atomic nuclei. In this respect, the problem of nuclear constitution exhibits a characteristic difference from that of the constitution of the extra-nuclear electron configuration, since the stability of this configuration is essentially independent of the mass-ratio. When we pass from the helium nucleus to heavier nuclei, the problem of nuclear constitution is, of course, still more complicated, although a certain simplification is afforded by the circumstance that the α-particles can be considered to a large extent to enter as separate entities into the constitution of these nuclei. This is not only suggested by the general facts of radioactivity, but appears also from the smallness of the additional mass defect, expressed by Aston's whole-number rule for the atomic weights of isotopes.

The main source of knowledge regarding the constitution of atomic nuclei is the study of their disintegrations, but important information is also derived from ordinary spectral analysis. As was mentioned, the hyperfine structures of spectral lines allow us to draw conclusions concerning the magnetic moments and angular momenta of the atomic nuclei, and from the intensity variations in band spectra we deduce the statistics obeyed by the nuclei. As might be expected, the interpretation of these results falls largely outside the scope of present quantum mechanics, and, in particular, the idea of spin is found not to be applicable to intra-nuclear electrons, as was first emphasized by Kronig. This situation appears especially clearly from the evidence concerning nuclear statistics. It is true that the fact, already mentioned, that the helium nuclei obey the Bose statistics is just what was to be expected from quantum mechanics for a system composed of an even number of particles which, like the electrons and protons, satisfy Pauli's exclusion principle. But the next nucleus for which data concerning statistics are available, namely the nitrogen nucleus, obeys also the Bose statistics, although it is composed of an uneven number of particles, namely 14 protons and 7 electrons, and thus should obey the Fermi statistics. Indeed, the general experimental evidence concerning this point seems to follow the rule that nuclei containing an even number of protons obey the Bose statistics, while nuclei containing an uneven number of protons obey the Fermi statistics. On the one hand, this remarkable "passivity" of the intra-nuclear electrons in the determination of the statistics is a very direct indication, indeed, of the essential limitation of the idea of separate dynamical entities when applied to electrons. Strictly speaking, we are not even justified in saying that a nucleus contains a definite number of electrons, but only that its negative electrification is equal to a whole number of elementary units, and, in this sense, the expulsion of a β-ray from a nucleus may be regarded as the creation of an electron as a mechanical entity. On the other hand, the rule just mentioned regarding nuclear statistics may be considered, from this point of view, as a support for the essential validity of a quantum mechanical treatment of the behaviour of the α-particles and protons in the nuclei. Actually, such a treatment has also been very fruitful in accounting for their part in spontaneous and controlled nuclear disintegrations.

In the ten years that have elapsed since Rutherford's fundamental discoveries, a large amount of most valuable material on this subject has been accumulated, owing, above all, to the great exploration work in the new field carried on in the Cavendish Laboratory under his guidance. Now, from the theoretical standpoint, it is one of the most interesting results of the recent development of atomic theory that the use of probability considerations in the formulation of the fundamental *disintegration law*, which for its time was a quite isolated and very bold hypothesis, has been found to fall entirely in line with the general ideas of quantum mechanics. Already at the more primitive stage of the quantum theory, this point was touched upon by Einstein in connexion with his formulation of the probability laws of elementary radiation processes, and was further stressed by

Rosseland in his fruitful work on inverse collisions. It is the wave-mechanical symbolism, however, which first offered the basis for a detailed interpretation of radio-active disintegrations, in complete conformity with Rutherford's deduction of nuclear dimensions from the scattering of α-rays. As was pointed out by Condon and Gurney, and independently by Gamow, the wave-formalism leads, in connexion with a simple model of the nucleus, to an instructive explanation of the law of α-ray disintegration as well as of the peculiar relationship, known as the rule of Geiger and Nuttall, between the mean life-time of the parent element and the energy of the α-ray expelled. Gamow, especially, succeeded in extending the quantum mechanical treatment of nuclear problems to a general qualitative account of the relationship between α- and γ-ray-spectra, in which the ideas of stationary states and elementary transition processes play the same part as in the case of ordinary atomic reactions and the emission of optical spectra. In these considerations, the α-particles in the nuclei are treated similarly to the extra-nuclear electrons in the atoms, with the characteristic difference, however, that the α-particles obey the Bose statistics and are kept within the nucleus by their own interaction, while the electrons, obeying the Fermi statistics, are held in the atom by the attraction of the nucleus. This is, among other causes, responsible for the smallness of the rate of energy emission, as γ-radiation, from excited nuclei which is even comparable with the rate of mechanical energy exchange between such nuclei and the surrounding electron clusters, the so-called internal conversion. In fact, in contrast to an atom built up of separate positive and negative particles, a nucleus-like system composed only of α-particles will never possess an electric moment, and, in this respect the additional protons and negative electrification of actual nuclei can hardly be expected to make much difference. Apart from such simple applications of the correspondence argument, our ignorance of the forces acting on the α-particles and protons in the nuclei, which must be assumed to depend essentially on the negative electrification, prevents at present theoretical predictions of a more quantitative character. A promising means of exploring these forces is afforded, however, by the study of controlled disintegrations and allied phenomena. As far as the behaviour of α-particles and protons is concerned, it may therefore be possible to build up gradually, by means of quantum mechanics, a detailed theory of nuclear constitution, from which in turn we may get further information about the new aspects of atomic theory presented by the problem of negative nuclear electrification.

As regards this last question, much theoretical interest has recently been aroused by the peculiar features exhibited by the *β-ray expulsions*. On the one hand, the parent elements have a definite rate of decay, expressed by a simple probability law, just as in the case of the α-ray disintegrations. On the other hand, the energy liberated in a single β-ray disintegration is found to vary within a wide continuous range, whereas the energy emitted in an α-ray disintegration, when due account is taken of the accompanying electromagnetic radiation and the mechanical energy conversion, appears to be the same for all atoms of the same element. Unless the expulsion of β-rays from atomic nuclei, contrary to expectation, is not a spontaneous process but caused by some external agency, the application of the principle of energy conservation to β-ray disintegrations would accordingly imply that the atoms of any given radio-element would have different energy contents. Although the corresponding variations in mass would be far too small to be detected by the present experimental methods, such definite energy differences between the individual atoms would be very difficult to reconcile with other atomic properties. In the first place, we find no analogy to such variations in the domain of non-radioactive elements. In fact, as far as the investigations of nuclear statistics go, the nuclei of any type, which have the same charge and, within the limits of experimental accuracy, the same mass, are found to obey definite statistics in the quantum mechanical sense, meaning that such nuclei are not to be regarded as approximately equal, but as essentially identical. This conclusion is the more important for our argument, because, in absence of any theory of the intra-nuclear electrons, the identity under consideration is in no way a consequence of quantum mechanics, like the identity of the extra-nuclear electronic configurations of all atoms of an element in a given stationary state, but represents a new fundamental feature of atomic stability. Secondly, no evidence of an energy variation of the kind in question can be found in the study of the stationary states of the radioactive nuclei involved in the emission of α- and γ-rays from members of a radioactive family preceding or following a β-ray product. Finally, the definite rate of decay, which is a common feature of α- and β-ray disintegrations, points, even for a β-ray product, to an essential similarity of all the parent atoms, in spite of the variation of the energy liberated by the expulsion of the β-ray. In absence of a general consistent theory embracing the relationship between the intrinsic stability of electrons and protons and the existence of the elementary quanta of electricity and action, it is very difficult to arrive at a definite conclusion in this matter. At the present stage of atomic theory, however, we may say that we have no argument, either empirical or theoretical, for upholding the energy principle in the case of β-ray disintegrations, and are even led to complications and difficulties in trying to do so. Of course, a radical departure from this principle would imply strange consequences, in case such a process could be reversed. Indeed, if, in a collision process, an electron could attach itself to a nucleus with loss of its mechanical individuality, and subsequently be recreated as a β-ray, we should find that the energy of this β-ray would generally differ from that of the original electron. Still, just as the account of those aspects of atomic constitution essential for the explanation of the ordinary physical and chemical properties of matter implies a renunciation of the classical ideal of causality, the features of atomic stability, still deeper-lying, responsible for the existence and the properties of atomic nuclei, may force us to renounce the very idea of energy balance. I shall not enter further into such speculations and their possible bearing on the much debated question of the source of stellar energy. I have touched upon them here mainly to emphasize that in atomic theory, notwithstanding all the recent progress, we must still be prepared for new surprises.

3
On the Structure of Atomic Nuclei. I†

W. HEISENBERG

ABSTRACT

We discuss the implications of our assumption that atomic nuclei are made up of protons and neutrons and do not contain electrons.

1. The Hamiltonian function of the nucleus.
2. The relation of charge and mass and the special stability of the He-nucleus.

3–5. The stability of nuclei and radioactive decay series.

6. Discussion of the physical assumptions.

THE experiments of Curie and Joliot‡ and their interpretation by Chadwick§ have shown that in the structure of nuclei a new, fundamental component, the neutron, plays an important part. This suggests that atomic nuclei are composed of protons and neutrons, but do not contain any electrons.|| If this is correct, it means a very considerable simplification of nuclear theory. The fundamental difficulties of the theory of β-decay and the statistics of the nitrogen-nucleus can then be reduced to the question: In what way can a neutron decay into proton and electron and what statistics does it satisfy? The structure of nuclei, however, can be described, according to the laws of quantum mechanics, in terms of the interaction between protons and neutrons.

† *Zeits. f. Phys.*, **77**, 1 (1932).
‡ I. Curie and F. Joliot, *C. R.* **194**, 273, 876 (1932).
§ J. Chadwick, *Nature*, **129**, 312 (1932).
|| Cf. also D. Iwanenko, *ibid.* p. 798.

1

We assume that neutrons obey the laws of Fermi statistics and have spin $\frac{1}{2}\hbar$. This assumption is necessary in order to explain the statistics of the nitrogen-nucleus and it is consistent with the empirical results on nuclear moments. If we wanted to describe the neutron as composed of proton and electron, the electron would have to obey Bose statistics and have spin zero; but we do not consider this any further. The neutron will be taken as an independent fundamental particle which, however, can split, under favourable conditions, into a proton and an electron, violating the law of conservation of energy and momentum.† We first investigate the interactions between a neutron and a proton. If a neutron and a proton are brought together to a distance comparable with nuclear dimensions, the negative charge changes place—analogous to the H_2^+-ion—with a frequency given by a function $(1/h)\ J(r)$ of the distance r between the two particles. The quantity $J(r)$ corresponds to the exchange-integral of the molecular theory. This can be illustrated by imagining the exchange of electrons without spin which obey the laws of Bose-statistics. However, it may be more correct to interpret the exchange-integral $J(r)$ as a fundamental property of the neutron–proton pair, without reducing it to the movements of electrons.

Similarly, we describe the interaction between two neutrons by the interaction-energy $-K(r)$, and we assume, in analogy to the H_2-molecule, that this energy leads to an attractive force between the neutrons. Finally, D denotes the mass-defect of the neutron relative to the proton (in energy units). We then assume that, apart from the forces given by functions $J(r)$ and $K(r)$ and by the Coulomb repulsion e^2/r between two protons, no more forces between the nuclear particles will appear. Also, we neglect all relativistic effects such as the spin-orbit interaction. Only a few, very general statements can be made about functions $J(r)$ and $K(r)$. We suspect that in the region of order 10^{-12} cm they decrease to zero with increasing r. Further, in analogy to molecules, we assume that for normal values of r the function $J(r) > K(r)$: this will be important later on. The mass-defect

† Cf. N. Bohr, Faraday Lecture, *J. Chem Soc.* 349 (1932).

D of the neutron should be small compared with the normal mass-defect of the elements.

In order to write down the Hamiltonian function of the nucleus the following variables are useful: Each particle in the nucleus is characterized by five quantities: the three position-coordinates $(x, y, z) = \mathbf{r}$, the spin σ^z along the z-axis and a fifth number ρ^ζ which can be ± 1. $\rho^\zeta = +1$ means the particle is a neutron, $\rho^\zeta = -1$ means the particle is a proton. Because of the exchange there appear in the Hamiltonian transitional elements changing $\rho^\zeta = +1$ into $\rho^\zeta = -1$ and it is useful to introduce the following matrices

$$\rho^\xi = \begin{pmatrix} 0 & 1 \\ 1 & 0 \end{pmatrix}, \quad \rho^\eta = \begin{pmatrix} 0 & -i \\ i & 0 \end{pmatrix}, \quad \rho^\zeta = \begin{pmatrix} 1 & 0 \\ 0 & -1 \end{pmatrix}$$

The space ξ, η, ζ is, of course, not the ordinary space.

In terms of these variables the complete Hamiltonian function of the nuclei (M proton-mass, $r_{kl} \rightarrow |\mathbf{r}_k - \mathbf{r}_l|$, $\mathbf{p}_k$ momentum of particle k) is

$$H = \frac{1}{2M} \sum_k \mathbf{p}_k^2 - \tfrac{1}{2} \sum J(r_{kl})(\rho_k^\xi \rho_l^\xi + \rho_k^\eta \rho_l^\eta) + \tfrac{1}{4} \sum_{k>l} K(r_{kl}).(1+\rho_k^\zeta)(1+\rho_l^\zeta) +$$
$$+ \tfrac{1}{4} \sum_{k>l} \frac{e^2}{r_{kl}} (1-\rho_k^\zeta)(1-\rho_l^\zeta) - \tfrac{1}{2} D \sum_k (1+\rho_k^\zeta) \quad (1)$$

Of the five terms the first is the kinetic energy of the particles, the second the exchange energy, the third the attractive forces between neutrons, the fourth the Coulomb repulsion of the protons and the fifth the mass-defects of the neutrons. It is now a purely mathematical problem to deduce the structure of nuclei from equation (1).

2

We now investigate a nucleus of n particles, i.e. n_1 neutrons and n_2 protons. $n_1 = \tfrac{1}{2} \sum_k (1 + \rho_k^\zeta)$ commutes with H in equation (1), i.e. it is a constant of the motion. The same holds for n_2. If we neglect the last three terms in equation (1) retaining only the two first, the energy for reasons of symmetry stays the same even if the sign of $\sum \rho_k^\zeta$ changes. For $\sum_k \rho_k^\zeta = 0$ we must have an extreme value of the energy. As there is no binding energy for $\sum_k \rho_k^\zeta = n$ with this assumption, the minimum value of the energy should in general correspond to $\sum_k \rho_k^\zeta = 0$. In other words the first two terms of the Hamiltonian are completely symmetrical with regard to protons and neutrons. The minimum energy obtainable through an exchange interaction occurs if the nucleus contains an equal number of neutrons and protons. This agrees with experimental results that the mass of nuclei is generally twice as big as their charge (in terms of units of charge and mass of the proton). The three last terms of equation (1) change the neutron/proton ratio corresponding to the minimum energy in favour of the neutrons, and increasingly so as the total number n grows bigger, because of the Coulomb forces of the protons. The application of this result to the problem of which nuclei can exist in nature and which cannot, requires a detailed discussion of nuclear stability and will only be treated in Sections 3–5.

The only nucleus for which the solution of equation (1) can be found immediately is Urey's† hydrogen isotope of weight 2. It consists of one proton and one neutron, and the wave-function $\psi(\mathbf{r}_1\rho_1^\zeta, \mathbf{r}_2\rho_2^\zeta)$ that solves equation (1), can always be written as (analogous to the He-problem of quantum mechanics)

$$\psi(\mathbf{r}_1\rho_1^\zeta, \mathbf{r}_2\rho_2^\zeta) = \phi(\mathbf{r}_1\mathbf{r}_2).(\alpha(\rho_1^\zeta)\beta(\rho_2^\zeta) \pm \alpha(\rho_2^\zeta)\beta(\rho_1^\zeta)) \tag{2}$$

where

$$\begin{aligned} \alpha(\rho) &= \delta_{\rho,1} \\ \beta(\rho) &= \delta_{\rho,-1} \end{aligned} \tag{3}$$

We have an attraction of the two particles, if we take the positive sign on the right hand side of equation (2). $\phi(\mathbf{r}_1, \mathbf{r}_2)$ then satisfies the wave equation

$$\left\{\frac{1}{2M}(\mathbf{p}_1^2 + \mathbf{p}_2^2) - J(r_{12}) - D - W\right\}\phi(\mathbf{r}_1\mathbf{r}_2) = 0 \tag{4}$$

In the lowest energy state $\phi(\mathbf{r}_1, \mathbf{r}_2)$ is symmetrical in $\mathbf{r}_1$ and $\mathbf{r}_2$. This is possible because of the spin despite the Fermi statistics of the particles.

† H. Urey, F. Brickwedde and G. Murphy, *Phys. Rev.* **39**, 164 (1932); **40**, 1, 464 (1932).

We shall not give a more detailed mathematical investigation of the He-nucleus according to equation (1) just now. Only the following qualitative ideas shall be discussed: If we investigate nuclei consisting of neutrons only, we find that a nucleus with two neutrons, according to equation (1), should be particularly stable, since the eigenfunction of the system may be symmetrical in two and only two neutrons (i.e. in its coordinates $\mathbf{r}$ and ρ) because of the Pauli principle. [The fact that nuclei containing only neutrons are unstable for reasons not contained in equation (1) will be discussed later and is of no importance for the following.] For the same reason we may assume that the He-nucleus with two protons and two neutrons acts as a closed shell according to the Pauli-principle, and it is thus particularly stable, as is shown by experiment. Also, its total spin vanishes.

We shall also investigate the interaction between two nuclei at a greater distance. We assume that for both nuclei $\sum \rho^\zeta = 0$, i.e. both nuclei contain the same number of protons and neutrons. The interaction between the nuclei, which can be taken as a small perturbation, is, according to equation (1)

$$H^{(1)} = -\tfrac{1}{2} \sum_{kk'} J(r_{kk'})(\rho_k^\xi \rho_{k'}^\xi + \rho_k^\eta \rho_{k'}^\eta) - \tfrac{1}{4} \sum_{kk'} K(r_{kk'})(1+\rho_k^\zeta)(1+\rho_{k'}^\zeta) + \\ + \tfrac{1}{4} \sum_{kk'} \frac{e^2}{r_{kk'}} (1-\rho_k^\zeta)(1-\rho_{k'}^\zeta) \quad (5)$$

where index k refers to the particles of one, index k' to the particles of the other nucleus. If we take the average time-value of equation (5) over the undisturbed motion of the nuclei, we are left with an average Coulomb repulsion of the nuclei and an average attraction of the neutrons; the repulsion predominating for great, the attraction for small distances. The average time-value of the biggest term in equation (5) vanishes, because the expected value of ρ^ξ vanishes, if $\sum \rho^\zeta = 0$ is known (which follows most simply from the symmetry of the problem in the ξ, η, ζ space along the ζ-axis). If, however, we carry the perturbation calculation to the second approximation, then the transition elements of the first term in equation (5) cause an attraction similar to van der Waals forces as the energy perturbation of the second order is always

$$W_k^{(2)} = -\sum_l \frac{|H_{kl}^{(1)}|^2}{h\nu_{kl}} \tag{6}$$

So, two nuclei at a great distance repell each other because of their charge, whereas, at a short distance they are bound by a van der Waals attraction and the attraction of the neutrons.

3

After what we have said, we may imagine the nucleus as a structure containing generally a few more neutrons than protons and in which two protons and two neutrons are bound together and form particularly stable particles, i.e. α-particles. Under what conditions is such a nucleus stable and in what way can it decay when there is instability?

Let us investigate first a nucleus containing only neutrons; because of the attraction of neutrons caused by the third term in equation (1) such a nucleus should be stable as it would be difficult to remove a neutron from it. But we would gain energy by removing a neutron from the nucleus and by adding a proton, since the gain in adding a proton is bigger than the loss in removing a neutron. This happens if we assume that the exchange forces dominate over the attractive forces between the neutrons. Thus we may assume that such a nucleus decays emitting β-rays. Although it is questionable, because of experimental results on continuous β-spectra, to apply the law of energy and momentum to the decay of a neutron we shall use it in some sort of way and state: β-decay takes place only if the rest mass of the nucleus investigated is bigger than the sum of the rest mass after β-decay and the rest mass of the electron. This assumption has been used before in the theory of the nucleus.† In explanation we may say that a neutron, analogous to quantum mechanical systems, would occasionally decay spontaneously under the influence of a strong electric field. If the total energy is positive it means that a field acts on the neutron in the nucleus and tries, like an electric

† G. Gamow, Constitution of Atomic Nuclei. Oxford (1931).

field, to decompose it. If the total energy (which is always accurately defined) is negative there is no such force.

Taking into account the above assumption on the stability of nuclei with regard to β-decay we may conclude: The nucleus consisting first of neutrons only will transform neutrons to protons emitting β-rays until the energy gained by adding a proton is exactly the same as the energy used to remove the neutron, i.e. until the minimum of the energy-curve with constant number of particles is reached. With smaller neutron numbers the nucleus is stable with regard to β-decay.

We can estimate the position of the minimum as a function of the mass-number as follows: Assuming that $J(r)$ vanishes rapidly as the distance increases the gain in exchange energy which is freed on adding a proton can essentially depend only on the neutron–proton ratio n_1/n_2; i.e. it is given by a function $f(n_1/n_2)$. Likewise for heavy nuclei the energy loss caused by removing a neutron will approach a value $g(n_1/n_2)$ which depends only on n_1/n_2. Finally, to add a proton we need to do work $n_2 e^2/R \simeq$ const. $n_2/n^{\frac{1}{3}}$ against electro-static forces, where R is the nuclear radius approximately proportional to $n^{\frac{1}{3}}$. The minimum thus is given by

$$f(n_1/n_2) = g(n_1/n_2) + \text{const } n_2/n^{\frac{1}{3}} \tag{7}$$

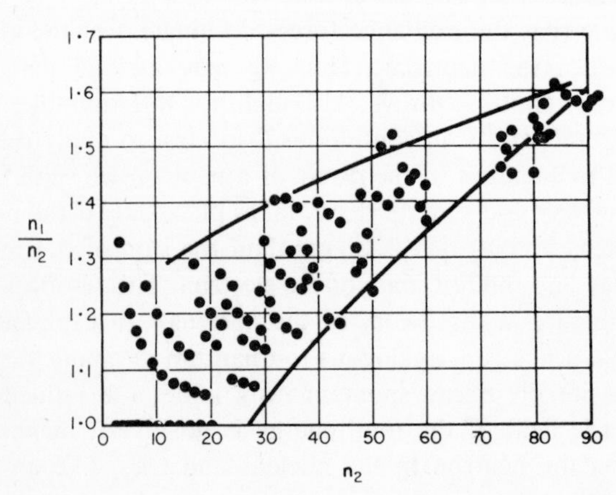

FIG. 1.

If we take $f(n_1/n_2)$ and $g(n_1/n_2)$ as approximately linear functions of n_1/n_2 we obtain the following approximation

$$n_1/n_2 = C_1 + C_2 n_2/n^{\frac{1}{3}} \tag{8}$$

where C_1 and C_2 are constant.

In Fig. (1) we give for each nuclear charge the maximum and minimum values of the ratio n_1/n_2 observed for the element in question. These values vary greatly. This can partly be explained by the fact that for many elements there may exist stable isotopes that have not yet been observed because of their rareness. For comparison with equation (8) we give a curve through the maxima of type (8) with constants $C_1 = 1{\cdot}173$, $C_2 = 0{\cdot}0225$. The qualitative aspect of the ratio n_1/n_2 in the nuclear system is well illustrated by a curve of type (8).

4

If the ratio n_1/n_2 reaches a certain critical value the Coulomb repulsion of the positive charges compared with the exchange and neutron forces can—particularly with heavy nuclei—become so big that the nucleus decays spontaneously emitting α-particles. As α-particles are generally much less strongly bound to the nucleus there is no emission of protons with this decay. Nuclei resulting from β-decay of higher nuclei could not emit protons as the β-decay always comes to an end when the removal of the proton still requires an energy supply.

The minimum value of the ratio n_1/n_2 follows from the condition that the Coulomb energy gained on emission of an α-particle is compensated by the other exchange energies of the α-particle in the remaining nucleus. In heavy nuclei these energies will depend only on the ratio n_1/n_2. If we assume again the dependence as approximately linear we obtain, as in equation (8), the equation

$$\frac{n_1}{n_2} = c_1 + c_2 \frac{n_2}{n^{\frac{1}{3}}} \tag{9}$$

Fig. (1) shows curve (9) with the constants $c_1 = 0{\cdot}47$, $c_2 = 0{\cdot}077$ representing approximately the position of the minima. On interpreting the two curves in Fig. (1) we must take into account that the four constants C_1, C_2, c_1, c_2 have been determined empirically, that equation (8) and (9) are approximations only, and, most important, that in a detailed theory the stability of a nucleus need not depend on the ratio n_1/n_2 only, but also on finer details of nuclear structure. As far as stability limits for α- and β-decay are concerned the two curves are only qualitatively important. In the region where the two curves are very close together we find the radioactive elements whose behaviour we shall discuss in more detail.

5

A superficial investigation of Fig. (1) shows that for radioactive elements the ratio n_1/n_2 alone is not sufficient to determine the stability of the nuclei. The critical ratio-numbers of the three radioactive families have different positions, and even within the radioactive decay series the stability towards β-decay depends on specific properties of the nucleus. Let us assume that at the beginning of a decay series there is a nucleus with an even number of protons and that it is stable towards β-decay. On emitting α-particles this nucleus will transform into nuclei with fewer protons and neutrons, and the ratio n_1/n_2 will increase until it exceeds a critical value. Then β-decay will set in, i.e. from the energy point of view it is easy to remove a neutron and add a proton; after this decay the proton number is odd. As the He-nucleus is very stable it is still possible, from the energy point of view, to transform a second neutron into a proton and thus build up a He-nucleus inside the nucleus. If the atomic number is even at the beginning the nucleus can always emit two β-particles, if it is odd only one β-particle is emitted. This law holds for all radioactive decay series. The critical ratio n_1/n_2 is higher for the emission of the first β-particle than for the second. After two β-particles have been emitted the ratio n_1/n_2 should generally be so low that β-decay stops. However, α-decay may follow and increase the ratio n_1/n_2 gradually until it exceeds a second time the critical value (for even proton number); then β-decay sets in again, etc. Eventually the nucleus becomes stable somewhere. It may also happen that a nucleus decays through both β- and α-decay; then we have the well known branchings which we shall not discuss here any further. Table (1) shows for the three radioactive decay-series the mass-number n_2, the neutron-number n_1 and the ratio n_1/n_2. The ratio-numbers for β-decay are in heavy print. The table shows that the second β-instability of the decay-series (in β-products) occurs at exactly the point where the ratio n_1/n_2 exceeds the critical value determined by the first β-instability. Only the third β-instability in the Ra-series (with RaD) cannot be explained by this simple idea.

TABLE 1

Thorium-series				Radium-series				Actinium-series			
Element	n_2	n_1	n_1/n_2	Element	n_2	n_1	n_1/n_2	Element	n_2	n_1	n_1/n_2
Th	90	142	1·579	U_1	92	146	1·588	Pa	91	144	1·582
α				α				α			
MTh_1	88	140	**1·591**	UX_1	90	144	**1·600**	Ac	89	142	**1·596**
β				β				β			
MTh_2	89	139	**1·562**	UX_2	91	143	**1·571**	RaAc	90	141	1·567
β				β				α			
RaTh	90	138	1·533	U_{11}	92	142	1·544	AcX	88	139	1·580
α				α				α			
ThX	88	136	1·545	Jo	90	140	1·556	AcEm	86	137	1·593
α				α				α			
ThEm	86	134	1·558	Ra	88	138	1·569	AcA	84	135	1·608
α				α				α			
ThA	84	132	1·571	RaEm	86	136	1·582	AcB	82	133	**1·622**
α				α				β			
ThB	82	130	**1·587**	RaA	84	134	1·595	AcC	83	132	**1·590**
β				α				β			
ThC	83	129	**1·555**	RaB	82	132	**1·610**	AcC′	84	131	1·560
β				β				α			
ThC′	84	128	1·524	RaC	83	131	**1·579**	AcD	82	129	1·573
α				β							
ThD	82	126	1·537	RaC′	84	130	1·548				
				α							
				RaD	82	128	**1·561**				
				β							
				RaE	83	127	**1·530**				
				β							
				RaF	84	126	1·500				
				α							
				RaG	82	124	1·512				

In the Th-series the critical ratios of the β-decay for even and odd proton numbers are approximately 1·585 and 1·55, in the Ra-series 1·595 and 1·57, in the Actinium-series 1·62 and 1·59. The β-decay of RaD, however, shows that apart from n_1/n_2 and the peculiar stability of the He-nucleus other structural properties of the nuclei can affect their stability.

6

To finish up with we shall briefly discuss the fundamental accuracy limits within which a Hamiltonian function of the nucleus of type (1) can describe the physical behaviour of nuclei. If we treat nuclei like molecules and thus compare neutrons with atoms it follows that equation (1) holds only if the motion of the protons is slow compared with the motion of the electron in the neutron; i.e. the speed of the proton must be small compared with the speed of light. For this reason we have left out all relativistic terms in the Hamiltonian (1). The mistake we thus make is of the order $(v/c)^2$, i.e. about 1 per cent. In this approximation the neutron can, so to speak, be taken as a static structure. However, we must realize that there are other physical phenomena where the neutron cannot be interpreted like this and for which equation (1) does not hold. Such phenomena are the Meitner-Hupfeld effect, the scattering of γ-rays on nuclei, further all experiments which split neutrons into protons and electrons (an example is the stopping of cosmic ray electrons on the passage through nuclei). For the discussion of such experiments we have to investigate more precisely all the fundamental difficulties that appear in the continuous β-ray spectra.

Note :

(i) In a later paper Heisenberg pointed out that the constants C_1 and C_2 in equation (8) did not correspond with the curve in Fig. 1. These constants should be corrected to $C_1=1{\cdot}16$ and $C_2=0{\cdot}0313$.

(ii) When discussing nuclear stability Heisenberg considered stability against β-decay and α-decay but not against positron decay. The positron was discovered after he wrote this paper.

(iii) Heisenberg believed that the internal structure of the neutron was important for the scattering of γ-rays by nuclei. We know now that this is not so. The secondary γ-rays observed by Meitner and Hupfeld were due to annihilation of positrons produced by primary γ-rays.

4

On the Structure of Atomic Nuclei. III†

W. HEISENBERG

THE experiments by Curie, Joliot and Chadwick on the existence and stability of the neutron led us to investigate in papers I and II of this series the part that neutrons play in the structure of nuclei, to propose certain physical assumptions and to test them with the facts of nuclear physics. The incomplete empirical results obtained to date cause a great uncertainty in any sort of theory, and only in very few instances do the experiments yield a definite interpretation. For this reason it seemed useful to formulate a certain hypothesis and investigate in what way it could put some order into the empirical results. In the following we shall also discuss in detail the consequences of our hypothesis and show where a different fundamental assumption would lead to the same results. We now supplement the ideas of the first two parts and clarify them in certain points.

1. The Thomas–Fermi Method with the Atomic Nucleus

In Part I we used a Hamiltonian function depending on the position coordinates $\mathbf{r}_k$ of the particles and the conjugate momenta $\mathbf{p}_k$, also on the variables ρ_k^ζ which indicate whether the particle is a neutron ($\rho_k^\zeta = +1$) or a proton ($\rho_k^\zeta = -1$). We also introduced in equation (1) the exchange integrals $J(r_{kl})$ and $K(r_{kl})$

† *Zeits. f. Physik.* **80**, 587 (1933).

and we shall now add, to complete the analogy with the molecular interactions, a "static" interaction $L(rkl)$ between neutrons and protons which corresponds to the electrostatic part of the binding energy of H and H^+ in the H_2^+-ion. In Part I we left out this term as we assumed it to be small. The complete Hamiltonian is now

$$H = \frac{1}{2M}\sum_k \mathbf{p}_k^2 - \tfrac{1}{2}\sum_{k>l} J(r_{kl})(\rho_k^\xi\rho_l^\xi + \rho_k^\eta\rho_l^\eta) + \tfrac{1}{2}\sum_{k>l} L(r_{kl})(1-\rho_k^\zeta\rho_l^\zeta) -$$
$$-\tfrac{1}{4}\sum_{k>l} K(r_{kl})(1+\rho_k^\zeta)(1+\rho_l^\zeta) + \tfrac{1}{4}\sum_{k>l}\frac{e^2}{r_{kl}}(1-\rho_k^\zeta)(1-\rho_l^\zeta) -$$
$$-\tfrac{1}{2}D\sum_k (1+\rho_k^\zeta) \qquad (1)$$

With many-particle nuclei we find an approximation of the solution of equation (1), analogous to the Thomas–Fermi method, in the following way: We first take the Schrödinger-function of the normal state in equation (1) as the solution of the variation problem

$$\int \psi^* H \psi \, d\Omega = \min \qquad (2)$$

with the condition

$$\int \psi^* \psi \, d\Omega = 1 \qquad (3)$$

If we admit in (2) only Schrödinger-functions with a definite numerical value of

$$4P(P+1) = \Big(\sum_k \rho_k^\xi\Big)^2 + \Big(\sum_k \rho_k^\eta\Big)^2 + \Big(\sum_k \rho_k^\zeta\Big)^2 \qquad (4)$$

(which is the total ρ-spin†) we find for $J(r_{kl}) > 0$ that $2P = n = n_1 + n_2$ leads to the lowest energy. ψ can be written in this approximation as

$$\psi(\mathbf{r}_1\rho_1^\zeta \ldots \mathbf{r}_n\rho_n^\zeta) = \varphi(\mathbf{r}_1 \ldots \mathbf{r}_n) f(\rho_1^\zeta \ldots \rho_n^\zeta) \qquad (5)$$

where f is a symmetrical function of ρ_k^ξ which can be calculated with the usual techniques of quantum mechanics provided we know $\sum_k \rho_k^\zeta = n_1 - n_2$.

The function φ then is a wave-function belonging to a Hamiltonian which we obtain from equation (1) by substituting the expressions

† Similarly we can, according to J. C. Slater, *Phys. Rev.* **35**, 210 (1930), interpret Hartree's method in atoms as an approximation of the variation problem which admits only a certain simple type of wave-function.

depending on ρ_k by their expected value $P = n/2$, $P^\zeta = \sum \rho_k^\zeta = n_1 - n_2$. For these expectation values we find

$$\left.\begin{aligned} k \neq l \quad \overline{\rho_k^\xi \rho_l^\xi + \rho_k^\eta \rho_l^\eta} &= 4\frac{n_1 n_2}{n(n-1)} \\ \overline{1 - \rho_k^\zeta \rho_l^\zeta} &= 4\frac{n_1 n_2}{n(n-1)} \\ \overline{(1+\rho_k^\zeta)(1+\rho_l^\zeta)} &= 4\frac{n_1(n_1-1)}{n(n-1)} \\ \overline{(1-\rho_k^\zeta)(1-\rho_l^\zeta)} &= 4\frac{n_2(n_2-1)}{n(n-1)} \\ \sum_k (1+\rho_k^\zeta) &= 2n_1 \end{aligned}\right\} \tag{6}$$

The modified Hamiltonian is therefore

$$H = \frac{1}{2M}\sum \mathbf{p}_k^2 - 2\frac{n_1 n_2}{n(n-1)}\sum_{k>l}[J(r_{kl}) - L(r_{kl})] - \frac{n_1(n_1-1)}{n(n-1)} \times \times \sum_{k>l} K(r_{kl}) + \frac{n_2(n_2-1)}{n(n-1)}\sum_{k>l}\frac{e^2}{r_{kl}} - n_1 D \tag{7}$$

The symmetry properties of $\varphi(\mathbf{r}_1 \ldots \mathbf{r}_n)$ with regard to the exchange of the particle coordinates are, as with atoms, prescribed by the Pauli-principle. The nucleus according to equation (7) appears to be a mechanical system of mass-points where the exchange energy of two mass-points is given by the expression

$$U(r) = -2\frac{n_1 n_2}{n(n-1)}[J(r) - L(r)] - \frac{n_1(n_1-1)}{n(n-1)} K(r) + + \frac{n_2(n_2-1)}{n(n+1)}\frac{e^2}{r} \tag{8}$$

If we interpret, like Thomas and Fermi, the nucleus as a gas of free particles that obey Fermi-statistics and are bound together by the forces of equation (8), and if $\rho(r)$ is the number of particles per unit of volume the kinetic energy of this gas is according to Fermi

$$E_{\text{kin}} = \frac{h^2}{M}\frac{4\pi}{5}\left(\frac{3}{8\pi}\right)^{\frac{5}{3}} \int \rho(\mathbf{r})^{\frac{5}{3}}\, d\tau \tag{9}$$

and the total energy of the nucleus

$$E = \frac{\boldsymbol{h}^2}{M}\frac{4\pi}{5}\left(\frac{3}{8\pi}\right)^{\frac{2}{3}}\int \rho(\mathbf{r})^{\frac{5}{3}}\,\mathrm{d}\tau + \\ +\tfrac{1}{2}\iint \rho(\mathbf{r})\rho(\mathbf{r}')U(|\mathbf{r}-\mathbf{r}'|)\,\mathrm{d}\tau\,\mathrm{d}\tau' - n_1 D \quad (10)$$

The density distribution $\rho(\mathbf{r})$ is determined by E becoming a minimum under the condition that

$$\int \rho(\mathbf{r})\,\mathrm{d}\tau = n$$

We have, however, to remember when using the Thomas–Fermi method that the approximation (10) for the energy is correct only under certain conditions. If, for instance, the function $U(|\mathbf{r}-\mathbf{r}'|)$, resembling the Gamow barrier, and for large n_1 and n_2 depending on n_1/n_2 only, suddenly increases considerably at a certain point, i.e. if very big repulsive forces hinder the attraction of two particles, the integral $\iint \rho(\mathbf{r})\,\rho(\mathbf{r}')\,U(|\mathbf{r}-\mathbf{r}'|)\,\mathrm{d}\tau\,\mathrm{d}\tau'$ would diverge or, at least, give incorrect values for the potential energy, since it never happens that two particles are closer than the critical distance. In this case we obtain a much better approximation by introducing, analogous to the constant "b" in van der Waal's equation, a minimum separation between two particles and also a maximum density ρ_0. We therefore put the potential energy function $U(|\mathbf{r}-\mathbf{r}'|)$ equal to zero for values of $|\mathbf{r}-\mathbf{r}'|$ smaller than the minimum separation between two particles. In the kinetic energy we have, exactly as in van der Waal's equation, $\rho(1/\rho-1/\rho_0)^{-\frac{2}{3}}$ instead of $\rho^{\frac{5}{3}}$ (i.e. the number of particles per cm^3: ρ multiplied by the mean energy $\rho^{\frac{2}{3}}$ of the particles). Instead of equation (10) we obtain a more general expression

$$E = \frac{\boldsymbol{h}^2}{M}\frac{4\pi}{5}\left(\frac{3}{8\pi}\right)^{\frac{2}{3}}\int \left(\frac{1}{\rho}-\frac{1}{\rho_0}\right)^{-\frac{2}{3}}\rho\,\mathrm{d}\tau + \\ +\tfrac{1}{2}\iint \rho(\mathbf{r})\rho(\mathbf{r}')U_0(|\mathbf{r}-\mathbf{r}'|)\,\mathrm{d}\tau\,\mathrm{d}\tau' - n_1 D \quad (11)$$

where U_0 replaces U in order to show that in $U_0(|\mathbf{r}-\mathbf{r}'|)$ we omit contributions for which $|\mathbf{r}-\mathbf{r}'|$ is smaller than the minimum separation. By a variation of ρ and using the condition $\int \rho\,\mathrm{d}\tau = n$ we obtain from equation (11)

$$\frac{\boldsymbol{h}^2}{M}\frac{4\pi}{3}\left(\frac{3}{8\pi}\right)^{\frac{5}{3}}\left(\frac{1}{\rho}-\frac{1}{\rho_0}\right)^{-\frac{5}{3}}\left(\frac{1}{\rho}-\frac{3}{5\rho_0}\right)+ \\ +\int \rho(\mathbf{r}')U_0(|\mathbf{r}'-\mathbf{r}|)\,\mathrm{d}\tau'-\lambda=0 \qquad (12)$$

If we multiply equation (12) by $\mathrm{d}\rho/\mathrm{d}n$ and integrate over $\mathrm{d}\tau$ we obtain from equations (11) and (12)

$$\lambda=\frac{\mathrm{d}E}{\mathrm{d}n} \qquad (13)$$

In this we assume $U_0(r)$ is independent of n. Equation (12) only holds in regions where $\rho \neq 0$. Outside these regions the system is completely determined by $\rho=0$. By multiplying equation (12) by $\frac{1}{2}\rho$ and integrating we get

$$\frac{\boldsymbol{h}^2}{M}\frac{2\pi}{3}\left(\frac{3}{8\pi}\right)^{\frac{5}{3}}\int\left(\frac{1}{\rho}-\frac{1}{\rho_0}\right)^{-\frac{2}{3}}\left(\rho+\frac{2}{5}\frac{\rho^2}{\rho_0-\rho}\right)\mathrm{d}\tau+ \\ +\frac{1}{2}\int\int \rho(\mathbf{r})\rho(\mathbf{r}')U_0\,\mathrm{d}\tau\,\mathrm{d}\tau'-\frac{n}{2}\frac{\mathrm{d}E}{\mathrm{d}n}=0 \qquad (14)$$

and by comparing with equation (11)

$$E-\frac{n}{2}\frac{\mathrm{d}E}{\mathrm{d}n}=\frac{\boldsymbol{h}^2}{M}\frac{2\pi}{15}\left(\frac{3}{8\pi}\right)^{\frac{5}{3}}\times \\ \times\left[\int\left(\frac{1}{\rho}-\frac{1}{\rho_0}\right)^{-\frac{2}{3}}\rho\,\mathrm{d}\tau-2\int\frac{\rho}{\rho_0}\left(\frac{1}{\rho}-\frac{1}{\rho_0}\right)^{-\frac{5}{3}}\mathrm{d}\tau\right] \qquad (15)$$

If we use this formula to discuss the dependence of the mass-defects on n_1 and n_2 it follows from Aston's measurements that the nuclei have a maximum density ρ_0 which must be of the same order of magnitude as the α-particle. If we can assume, as in the usual Thomas–Fermi method, that $1/\rho_0 \ll 1/\rho$ the right side of equation (15) is positive and thus $-E$ as a function of n would have to increase more rapidly than const. n^2. Empirically, however, for small n, $-E$ is almost proportional to n, and for large n it increases even more slowly. Thus the function $U(|\mathbf{r}-\mathbf{r}'|)$ must increase rapidly for small separations $|\mathbf{r}-\mathbf{r}'|$. For heavy nuclei the density of the nucleus is near ρ_0, and at the surface, decreases to zero. Thus, if $J(r_{kl})$, $K(r_{kl})$,

$L(r_{kl})$ decrease rapidly with distance we obtain for the energy according to equation (15) for big n (n_1/n_2 = const.)

$$E = -an + bn^{\frac{5}{3}} + c \tag{16}$$

where $-an$ comes from the short-range forces, $bn^{\frac{5}{3}}$ from the Coulomb forces. We assumed this implicitly in Part I when discussing the stability curves.

It follows from the symmetry of the problem that $\rho(r)$ is spherically symmetrical, and it should not be too difficult to find approximate solutions for $\rho(r)$ if we know $U(r)$. For the time being we try to find the function $U(r)$ from the empirical mass-defects

5

On Nuclear Theory†

Ettore Majorana

Abstract

We discuss a new interpretation of Heisenberg's nuclear theory which leads to a slightly different Hamiltonian function. Accordingly we treat the nuclei statistically.

The discovery of the neutron, a heavy and uncharged elementary particle, made it possible to develop a nuclear theory using ideas of quantum mechanics without, however, removing the fundamental difficulties that are connected with β-decay. According to Heisenberg,‡ we can think of nuclei for many purposes as consisting of protons and neutrons, i.e. of particles of almost the same mass, with spin $\frac{1}{2}\hbar$ and obeying Fermi statistics. The problem is thus reduced to finding a suitable Hamiltonian which holds for this system of particles, and we need a non-relativistic approximation since the speed of the particles is presumably rather small compared with the speed of light ($v \sim c/10$). In order to find a suitable interaction between the components of the nuclei Heisenberg was guided by an obvious analogy. He treats the neutron as a combination of a proton and an electron, i.e. like a hydrogen atom bound by a process not fully understood by present theories, in such a way that it changes its statistical properties and its spin. He further assumes that there are exchange forces between protons and neutrons similar to those responsible for the molecular binding of H and H^+. In addition to

† *Zeits. f. Phys.* **82,** 137 (1933).

‡ W. Heisenberg, *Z. Phys.* **77,** 1 (1932) (see p. 144 of this book); **78,** 156 (1933).

this interaction between protons and neutrons considered essential for the stability of the nucleus, there are Coulomb repulsion between protons, van der Waals attraction between neutrons and some sort of electrostatic interaction between protons and neutrons.†

One may doubt the validity of this analogy as the theory does not explain the inner structure of the neutron, and the interaction between neutron and proton seems rather big compared with the mass-defect of the neutron as determined by Chadwick. We think, therefore, that it may be quite interesting to find a Hamiltonian very similar to Heisenberg's which represents in the simplest way the most general and most obvious properties of the nucleus. We shall use a statistical method which should be permissable for determining orders of magnitude. We should also like to point out the exchange forces must have the opposite sign to Heisenberg's forces because of the criterium we fixed for the Hamiltonian. Therefore, the symmetry characteristics of the eigenfunctions belonging to the normal state and the whole statistical treatment are different from Heisenberg's.

1

The numerous sources of information we have on nuclear structure, i.e. radioactive decay, artificial decay, anomalous scattering of α-particles, mass-defect measurements etc. seem to indicate that nuclei, unlike atoms, are not uniformly organized. On the contrary, it looks as though nuclei consist of rather independent components which react only on immediate contact, i.e. of some sort of matter with the same properties of size and impenetrability as macroscopic matter. Light and heavy nuclei are built up of this matter and the difference between them depends mainly on their different content of "nuclear matter". This theory can only be correct if the Coulomb repulsion between the positive components of nuclei is not very important. This certainly holds for rather light nuclei, whereas we have to have a slight correction for heavy nuclei.

If we assume that nuclei consist of protons and neutrons we have to formulate the simplest law of interaction between them which will

† W. Heisenberg, *Z. Phys.* **80**, 587 (1933) (see p. 155 of this book).

lead, if the electrostatic repulsion is negligible, to a constant density for nuclear matter. We have to find three laws of interaction: One between protons, one between protons and neutrons and one between neutrons. We shall assume, however, that only Coulomb's force acts between each pair of protons. This can be justified to a certain extent by the fact that the classical radius of protons is much smaller than the average distance between the particles in the nucleus. Also, the Coulomb force is not very important for light nuclei and, as they contain almost the same number of neutrons and protons, it seems reasonable to think that a special interaction between protons and neutrons is the main cause of nuclear stability. We assume that there is no noticeable interaction between the neutrons for there is

FIG. 1. Potential energy between two atoms.

no proof of the contrary. We now have to find a suitable interaction between protons and neutrons. Nuclear structure and the structure of solids and liquids seem to be somewhat similar and it might be possible to have an interaction of the same type as between atoms and molecules, i.e. attraction for large distances and strong repulsion for small distances so that the particles do not penetrate each other. We would also have to assume repulsive forces between neutrons with small separations in order to obtain the desired ratio between the number of particles and the nuclear volume. Such a solution would be aesthetically unsatisfactory, however, since we would have not only attractive forces of unknown origin between the particles, but also, for short distances, repulsive forces of enormous magnitude corresponding to a potential of several million volts. We shall, therefore, try to find another solution and introduce as few arbitrary elements as possible. The main problem is this: How can we obtain a density independent of the nuclear mass without obstructing the free movement of the particles by an artificial impenetrability? We must try to find an interaction whose average energy per particle never exceeds a certain limit however great the density. This might occur through a sort of saturation phenomenon more or less analogous to valence saturation. Such an interaction is given, as we shall prove, by

$$(Q', q'|J|Q'', q'') = -\delta(q' - Q'')\delta(q'' - Q')J(r) \tag{1}$$

where $r = |q' - Q'|$ and Q and q are the coordinates of a neutron and proton respectively. The function $J(r)$ is positive, and a possible form of it is shown in Fig. (2). Expression (1) implies that there is an

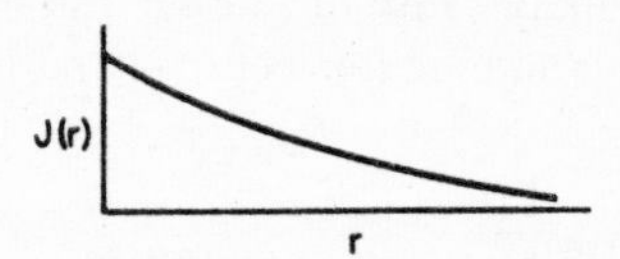

FIG. 2. Curve of the resonance forces.

attraction or a repulsion respectively between the neutron and the proton depending on whether the wave-function is approximately symmetrical or anti-symmetrical in the two particles. In order to account for the special stability of the α-particle we shall assume that Q and q in equation (1) are only the position coordinates without spin. Thus we find that both neutrons act on each proton in the α-particle instead of only one and vice versa, since we assume a symmetrical function in the position coordinates of all protons and neutrons (which is true only if we neglect the Coulomb energy of the protons). In the α-particle all four particles are in the same state so that it is a closed shell. If we proceed from an α-particle to heavier nuclei we can have no more particles in the same state because of the Pauli-principle. Also, the exchange energy (1) is usually large only if a proton and a neutron are in the same state and we may expect, which agrees with experiments, that in heavy nuclei the mass defect per particle is not noticeably bigger than in the α-particle.

Let us now compare expression (1) for the interaction between a proton and a neutron with the interaction deduced from the resonance term of Heisenberg's Hamiltonian by distinguishing between neutrons and protons and by eliminating the troublesome ρ-spin-coordinate.

We find an expression similar to (1) which is, however, fundamentally different in two respects: Firstly, in Heisenberg's expression Q and q stand for all coordinates including the spin. Secondly, Heisenberg assumes the opposite sign for the resonance forces. Statistically this is most important as there is no saturation because of the symmetry character of Heisenberg's eigenfunctions and repulsive interactions at short distances are necessary.† We shall now investigate the saturation that leads to the uniform density of the nuclear components found experimentally.

2

In a first approximation we take the eigenfunction of the nucleus as a product of two functions which depend on the coordinates of the n_1 neutrons and n_2 protons respectively:

$$\psi = \psi_N(Q_1, \Sigma_1, \ldots, Q_{n_1}, \Sigma_{n_1})\psi_P(q_1, \sigma_1, \ldots, q_{n_2}, \sigma_{n_2}) \qquad (2)$$

and we assume that ψ_N and ψ_P can be obtained by anti-symmetrizing products of individual orthogonal single-particle eigenfunctions:

$$\left.\begin{aligned} \psi_N &= \frac{1}{(n_1\,!)^{\frac{1}{2}}}\sum_R \pm R\psi'_N(Q_1, \Sigma_1)\ldots\psi_N^{n_1}(Q_{n_1}, \Sigma_{n_1}) \\ \psi_P &= \frac{1}{(n_2\,!)^{\frac{1}{2}}}\sum_R \pm R\psi'_P(q_1, \sigma_1)\ldots\psi_P^{n_2}(q_{n_2}, \sigma_{n_2}) \end{aligned}\right\} \qquad (3)$$

For many particles the individual-particle wave-function ψ may be identified with free-particle wave-packets. We can show that each proton is subject on the average to the interaction of a small number (one or two) of neutrons and vice versa, and the assumption of free-particle wave-functions introduces a slight error because of polarization effects. This method is, however, suitable for order-of-magnitude calculations.

† W. Heisenberg, *Z. Phys.* **80**, 587 (1933) (see p. 155 of this book). I would like to thank Professor Heisenberg very much for being able to see his paper before it was published.

We have to calculate the mean value of the total energy using the wave-function (2) and find its minimum. The energy consists of three parts:

$$W = T+E+A \tag{4}$$

where T is the kinetic energy, E the electrostatic energy of the protons and A the exchange energy. We assume that all individual particle states are either free or occupied twice with opposite spin direction. Then, n_1 and n_2 are even. We also introduce Dirac's density matrices:

$$\left.\begin{aligned}(q'|\rho_N|q'') &= \sum_{\sigma_i=1}^{2}\sum_{i=1}^{n_1}\psi_N^i(q',\sigma_i)\bar{\psi}_N^i(q'',\sigma_i)\\(q'|\rho_P|q'') &= \sum_{\sigma_i=1}^{2}\sum_{i=1}^{n_2}\psi_P^i(q',\sigma_i)\bar{\psi}_P^i(q'',\sigma_i)\end{aligned}\right\} \tag{5}$$

and have

$$\rho_N^2 = 2\rho_N, \qquad \rho_P^2 = 2\rho_P \tag{6}$$

where the factor 2 comes from the spin. The eigenvalues of the density matrices are

$$\rho_N = \left\langle{}^2_0\right., \qquad \rho_P = \left\langle{}^2_0\right. \tag{7}$$

If the mass M of each particle is approximately the same for neutrons and protons we obtain

$$T = \frac{1}{2M}\,\mathrm{Tr}\,[(\rho_N+\rho_P)p^2] \tag{8}$$

$$E = \frac{e^2}{2}\int(q'|\rho_P|q')\frac{1}{|q'-q''|}(q''|\rho_P|q'')\,\mathrm{d}q'\,\mathrm{d}q''+\ldots \tag{9}$$

In equation (9) we have left out a term which is essentially the Coulomb exchange energy of the protons. This term has been calculated by Dirac† and is not very important when there are many particles. Finally we obtain:

$$A = -\int(q'|\rho_N|q'')J|q'-q''|(q''|\rho_P|q')\,\mathrm{d}q'\,\mathrm{d}q'' \tag{10}$$

If the number of particles is large ρ_N and ρ_P are almost diagonal matrices and even classical functions of p and q. The best relation

† P. A. M. Dirac, *Proc. Camb. phil. Soc.* **26**, 376 (1930).

between the matrices and the classical functions is given by

$$\left.\begin{aligned}\left(q-\frac{v}{2}\middle|\rho_N\middle|q+\frac{v}{2}\right)&=\frac{1}{\boldsymbol{h}^3}\int\rho_N(p,q)\exp\left[-(2\pi i/\boldsymbol{h})(p,v)\right]\mathrm{d}p\\ \left(q-\frac{v}{2}\middle|\rho_P\middle|q+\frac{v}{2}\right)&=\frac{1}{\boldsymbol{h}^3}\int\rho_P(p,q)\exp\left[-(2\pi i/\boldsymbol{h})(p,v)\right]\mathrm{d}p\end{aligned}\right\}\qquad(11)$$

and by an inversion of the Fourier integrals. If we put equation (11) in the above expression we obtain

$$T=\frac{1}{2M}\int\frac{\rho_N(p,q)+\rho_P(p,q)}{\boldsymbol{h}^3}p^2\,\mathrm{d}p\,\mathrm{d}q\qquad(12)$$

$$E=\frac{e^2}{2}\int\frac{\rho_P(p,q)\rho_P(p',q')}{\boldsymbol{h}^6}\frac{1}{|q-q'|}\mathrm{d}p\,\mathrm{d}q\,\mathrm{d}p'\,\mathrm{d}q'\qquad(13)$$

$$A=\int\frac{\rho_N(p,q)V_N(p,q)}{\boldsymbol{h}^3}\mathrm{d}p\,\mathrm{d}q=\int\frac{\rho_P(p,q)V_P(p,q)}{\boldsymbol{h}^3}\mathrm{d}p\,\mathrm{d}q\qquad(14)$$

where $V_N(p,q)$ and $V_P(p,q)$ are the classical functions corresponding to the matrices

$$\left.\begin{aligned}(q'|V_N|q'')&=-(q'|\rho_P|q'')J\,|q'-q''|\\ (q'|V_P|q'')&=-(q'|\rho_N|q'')J\,|q'-q''|\end{aligned}\right\}\qquad(15)$$

We now assume that near a point q the states of low energy are occupied by neutrons as well as by protons. There will be a maximum value of the momentum $P_N(q)$ for the neutrons and the protons, and from equation (7) it follows that

$$\rho_N(p,q)=\left\langle\begin{aligned}&2,\ \text{if } p<P_N(q)\\ &0,\ \text{if } p>P_N(q)\end{aligned}\right\}\qquad(16)$$

$$\rho_P(p,q)=\left\langle\begin{aligned}&2,\ \text{if } p<P_P(q)\\ &0,\ \text{if } p>P_P(q)\end{aligned}\right\}\qquad(17)$$

We first investigate a limiting case, i.e. a case of very high density when h/p_N and h/p_P, which are of the order of magnitude of the mutual distance between the particles in the nucleus, are small compared with the range of the resonance forces. We also assume that $P_N>P_P$, i.e. that the density of the neutrons is larger than the density of the protons. We note that in the second equation of (15) ρ_N is almost diagonal and $J|q'-q''|$ can be substituted by $J(0)$ if

$J(0)$ is finite. The equation then reads

$$(q'|V_P|q'') = -J(0)(q'|\rho_N|q'')$$

and from this follows

$$V_P(p, q) = -J(0)\rho_N(p, q) \tag{18}$$

We put this in equation (14) and note that $\rho_N = 2$ if $\rho_P(p,q) > 0$ and obtain

$$A = -2J(0)\int \frac{\rho_P(p, q)}{h^3}\,\mathrm{d}p\,\mathrm{d}q = -2J(0)n_2 \tag{19}$$

This means that the binding energy per proton due to the exchange forces is only $-2J(0)$ if the particle density is high and the density of neutrons larger than that of protons. We neglect for the time being the Coulomb repulsion of the protons (which is approximately true for light nuclei) and fix the ratio n_1/n_2 but not the density. Then the potential energy per particle is a certain function of the total density

$$a = a(\mu), \qquad \mu = \frac{8\pi}{3h^3}(P_N^3 + P_P^3) \tag{20}$$

This vanishes for $\mu = 0$ and approaches a constant value $-\dfrac{2n_2}{n_1 + n_2}J(0)$ for $\mu \to \infty$. This limiting value will reach the minimum $-J(0)$ if $n_1 = n_2$. For intermediate densities the general expression of $a(\mu)$ follows from equations (10) and (11) and is

$$a = \frac{1}{\mu(q)}\int\int \frac{\rho_N(p, q)\rho_P(p', q)}{h^6}\,G(p, p')\,\mathrm{d}p\,\mathrm{d}p' \tag{21}$$

where $G(p, p')$ is a function of $|p-p'|$ which depends on $J(r)$ in the following way

$$G(p, p') = \int \exp\left[-(2\pi i/h)(p-p', v)\right] J(v)\,\mathrm{d}v \tag{22}$$

The kinetic energy per particle is

$$t = \kappa\mu^{\frac{2}{3}}$$

and the total energy $a+t$ reaches a minimum for a certain value dependent only on the ratio n_1/n_2 (Fig. 3). We obtain a constant density independent of the nuclear mass and thus a nuclear volume and binding energy proportional only to the number of particles, as is found by experiment. We can try to determine the function $J(r)$ in a way that best represents the experimental results. The expression

$$J(r) = \lambda \frac{e^2}{r}$$

for instance, with an arbitrary constant is suitable even though it becomes infinite if $r = 0$. For great distances, however, it must be

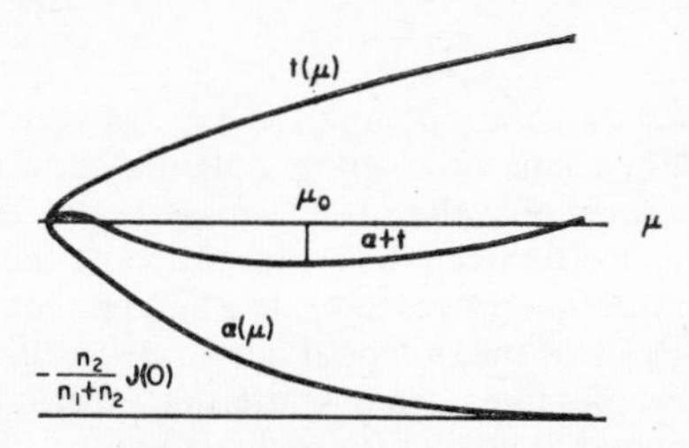

FIG. 3. Kinetic and potential energy per particle.

modified as it gives an infinite cross section for the collision between protons and neutrons. Also, it seems to provide too small a ratio for the mass defects of the α-particle and the hydrogen isotope. Thus, we have to use an expression with at least two constants, e.g. an exponential function, $J(r) = A\exp(-\beta r)$. We shall not follow this up since it has been shown that the first statistical approximation can lead to considerable errors however large the number of particles. For heavy nuclei Coulomb's force is very important which means that the nuclear extension increases slightly and the density of neutrons and protons is no longer constant locally. The exchange binding energy not only depends on the ratio n_1/n_2, it is even slightly smaller than for light nuclei because of the smaller density caused by Coulomb forces.

I would like to thank Professor Heisenberg very much for his advice and numerous discussions. My thanks are due to Professor Ehrenfest for many valuable discussions and also to the Consiglio Nazionale delle Ricerche for making my stay in Leipzig possible.

6

On the Mass Defect of Helium†

E. WIGNER

If one assumes that the potential energy between protons and neutrons has the shape of a simple potential hole, it is possible from the experimental value of the mass defect of the H^2, to derive a connection between the mean width and the depth of this curve. This connection proves to be, to a large extent, independent of the finer details of the potential curve. By assuming a certain probable value, obtained from scattering experiments, for the width of the potential hole, it is possible to make calculations on the mass defects of other nuclei. Such computations were carried out for He and yield values which are greater than the mass defect of H^2 by a rather large factor. This agrees with experiment. For the higher elements, the Pauli principle has to be taken into account and the structure of higher nuclei is discussed on this basis.

1

THE discovery of the neutron by Chadwick,‡ and by Curie and Joliot§ has made possible a more detailed picture of the constitution of the nuclei. As far as can be seen at present, there are three different assumptions possible concerning the elementary particles.

(a) The only elementary particles are the proton and the electron. This point of view has been emphasized by Heisenberg and treated by him in a series of papers.||

† *Phys. Rev.* **43**, 252 (1933).
‡ Chadwick, *Nature*, **129**, 469 (1932).
§ Curie and Joliot, *Comptes Rendus*, **193**, 1412 and 1415 (1931).
|| W. Heisenberg, *Z. Phys.* **77**, 1 (1932) (see p. 144 of this book); **78**, 156 (1932).

(b) The neutrons are elementary particles and the nuclei are built up by protons, electrons and neutrons. This point of view was proposed by Dirac and adopted by Bartlett† in his discussion of the constituents of the light elements.

(c) It may be assumed furthermore that in addition to the neutrons, discovered by Chadwick ("heavy neutrons") there are "light neutrons" of electronic mass, as first proposed by Pauli.‡ The number of light neutrons should be equal to the number of electrons in every nucleus and they leave the nucleus simultaneously with the β-rays. The number of electrons (and light neutrons) should be, just as in (b), equal to the number of "free electrons," as proposed by Beck.§ Some arguments in favour of this assumption were given by the present author.||

For the present purpose (the comparison of the mass-defects of the first few elements) it does not make any difference whether we adopt the hypothesis contained in (b) or (c), because the first elements, even up to Cl, do not contain any free electrons. The calculations will probably hold, even if the hypothesis (a) is adopted.

There seem to be three alternative possible assumptions concerning the nature of the forces acting between protons and neutrons. (The forces between two protons or between two neutrons are always neglected.) Heisenberg assumed that these forces are of the exchange type, similar to those of the H_2^+ molecule. If we suppose, however, that the neutrons have to be treated as elementary particles, one must either assume a certain potential energy $V(r)$ between a proton and a neutron, or a three-body force. The present calculations will be made on the basis of the former assumption. The other possibility is to calculate with a potential energy which is a function of the mutual distance of *three* particles. Forces of this kind¶ must be

† Cf. W. Bartlett, *Phys. Rev.* **42,** 145 (1932) (Letter to the Editor).
‡ Cf. Carlson and Oppenheimer, *Phys. Rev.* **38,** 1787 (1931).
§ G. Beck, *Z. Phys.* **47,** 407 (1928); **50,** 548 (1928).
|| E. Wigner, *Proc. Hung. Acad.* (1932) (to be sent to Pergamon).
¶ An example for such a potential is $cE^2(1+e^{r/\rho})^{-1}$, where c is a constant, r the distance of the neutron from one proton and E the electric field strength produced by the other protons.

assumed in the hypothesis (c) for the light neutrons, so it does not seem unnatural to allow them for the heavy neutrons as well.

The effect of the first kind of forces was fully discussed by Heisenberg and the discussion of the effect of the forces of the second kind can be carried out in a very similar way. One interesting feature of the second kind of forces is that it is probable that if a nucleus with n_p protons and n_h neutrons is stable and if n_p is odd then there is also a stable nucleus with n_p+1 protons and n_h neutrons. Also if n_h is odd there probably exists a stable nucleus with n_p protons and n_h+1 neutrons. The reason for this is that if n_p is odd, the next proton may have the same wave-function as this one, which is in conflict with the Pauli principle if n_p is even. From O up to Cl the nuclei predicted in this way are all known. Below oxygen, however, there are some nuclei lacking, namely those with the (n_p, n_h) values (1,2), (2,1), (4,3), (4,6), (6,5), (6,8), (8,7). A possible reason that these nuclei have so far escaped detection, together with a more exact proof of the above-mentioned rule, will be given in Section 3, a different explanation of their constitution was put forward by Jones.†

It may be seen furthermore, that just as in the theory of Heisenberg, the energies of the nuclei (n, n') and (n', n) are equal. Consequently among all nuclei with the same mass $n+n'$ that with the charge $n_p = (n_p+n_h)/2$ will be the most stable, having the largest number $(n_p+n_h)^2/4$ of attracting terms. The formation of the nuclei after O^{16} may be imagined like this†: Assuming that the addition of a heavy neutron to O^{16} is connected with an energy gain, we get O^{17}. Then according to the preceding rule, the capture of another neutron is possible, giving O^{18}. By this process the number of neutrons is increased so much in the nucleus, that it may capture a new proton giving F and then another, giving Ne^{20}. Now by the increased number of protons the capture of a new neutron is possible, giving Ne^{21}, and with another one Ne^{22}, and so on.

In addition to the difficulty connected with the apparent non-existence of the above-mentioned nuclei, it seems rather surprising that the nuclei between O and Cl adhere so very closely to the condi-

† E. G. Jones, *Nature*, **130**, 580 (1932).

tion $n_p = n_h$. This difficulty can be avoided, of course, by assuming a repulsive force between the neutrons and between the protons at small distances.

The three-body potential suggested on page 171 is also capable of explaining the qualitative features of the series of existing elements in some respects even better than that just discussed. It does not seen, however, to be easy to make simple assumptions as to the general shape of such a potential.

2

One of the remarkable facts about the mass defects in the very first elements is the very great binding energy of the He nucleus. The binding energy of the H^2 nucleus is only† three times the rest energy mc^2 of the electron, the binding energy of the He is‡ $52mc^2$, if we assume the mass of the neutron equal to the mass of the proton 1·00724 (referred to the mass of neutral O^{16}). The masses of the H^2 and He nuclei are taken to be 2·01297 and 4·00108, respectively. The binding energy of He is around 17 times larger than that of H^2.

This would rather indicate an attraction between the neutrons or between the protons, which is very unlikely on the basis of the previous discussion. The purpose of the subsequent calculation is to see how far it is possible to explain the large mass defect of He without such an assumption, or even to reconcile it with the existence of some repulsive forces between the different neutrons and also the different protons.

First we consider the H^2 nucleus. There are several indications that the first energy value depends only on the rough shape of the potential curve. For the H^2 nucleus, therefore, the potential energy was assumed for the purpose of the calculation to be

$$V(r) = 4v_0/(1+ \exp[r/\rho])(1+ \exp[-r/\rho]) \tag{1}$$

in units of mc^2, where v_0 and ρ are constants. The Schrödinger

† K. T. Bainbridge, *Phys. Rev.* **42,** 1 (1932). J. D. Hardy, E. F. Barker and D. U. Dennison, *Phys. Rev.* **42,** 279 (1932).
‡ F. W. Aston, *Proc. roy. Soc.* **A115,** 502 (1927).

equation becomes in this case

$$\left[-10\left(\frac{\partial^2}{\partial x^2}+\frac{\partial^2}{\partial y^2}+\frac{\partial^2}{\partial z^2}\right)+V\right]\psi(xyz)=\varepsilon\psi(xyz) \tag{2}$$

where x, y, z are the components of the distance between the two particles, and the energy and distance are always measured in units mc^2 and e^2/mc^2, respectively, and we have for convenience set $h^2mc^2/4\pi^2M = 10$. The characteristic numbers and functions of (2) with the potential (1) are known from the work of Eckart.† The lowest energy level is

$$-\varepsilon = 50/8\rho^2+v_0-(30/8\rho^2)(1+8v_0\rho^2/5)^{\frac{1}{2}} \tag{3}$$

while the corresponding (unnormalized) characteristic function is

$$\psi = \frac{\rho}{r}\frac{\exp[r/\rho]-1}{\exp[r/\rho]+1}\frac{1}{(1+\exp[r/\rho])^{\nu}(1+\exp[-r/\rho])^{\nu}} \tag{4}$$

with $\nu = (-\varepsilon\rho_2/10)^{\frac{1}{2}}$. The function $V(r)$ is graphically given in Fig. 1 (line 1). The constants ρ and v_0 must be chosen such that $\varepsilon = -3$ should give the observed binding energy of H^2. This gives an equation between v_0, the potential for $r = 0$, and ρ, the mean thickness of the potential hole, which is given in Fig. 2 (upper line). In order to have a better insight into the conditions governing the behaviour of the characteristic values and characteristic function, the characteristic function (4) for $\rho = 0{\cdot}22$, $v_0 = 140$ is given by the broken line in Fig. 1. One sees that it extends over a much wider region than $V(r)$ and in consequence the mean potential energy is much smaller than v_0. In Fig. 3, the mean negative potential energy $-P$, the mean kinetic energy K and the negative total energy $-\varepsilon = 3$ are plotted against the parameter ρ for the case in which v_0 is taken from Fig. 2, yielding (by equation (3)) $\varepsilon = -3$. For small values of ρ, the negative mean potential energy is much larger than $-\varepsilon$, and is almost totally compensated by the kinetic energy. Thus the value of ε is very sensitive to small variations of v_0, because these latter increase the mean potential energy without affecting the kinetic.

† C. Eckart, *Phys. Rev.* **35**, 1303 (1930).

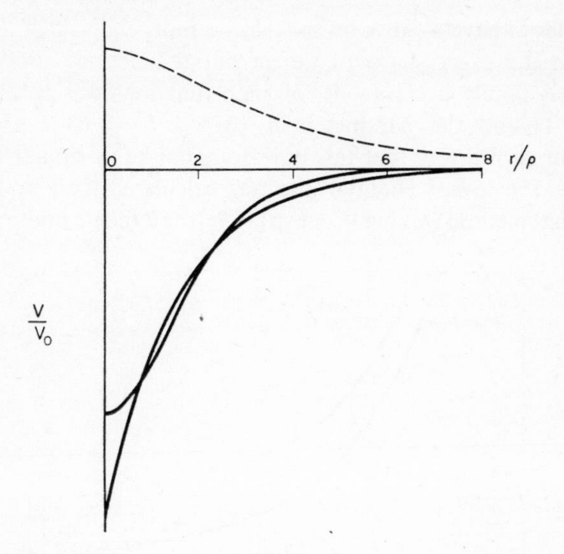

FIG. 1

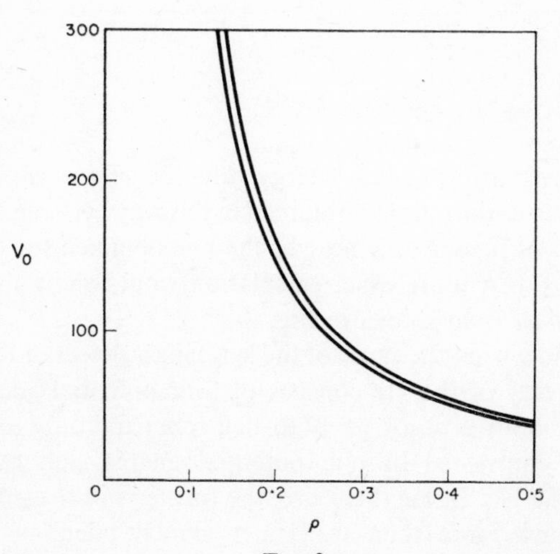

FIG. 2

In order to have a check on the relative independence of the (v_0, ρ) curve on the exact shape of the potential function, another two-parameter family $a \exp(-br)$ of such functions was taken (line 2 in Fig. 1) and the parameters $a = 1{\cdot}4v_0$, $b = 0{\cdot}63/\rho$ have been chosen in such a way that this new potential be as similar to (1) as possible. The lowest energy-value was calculated then by a simple variational method (taking $\psi = \exp[-\beta r]$) and then a and b adjusted

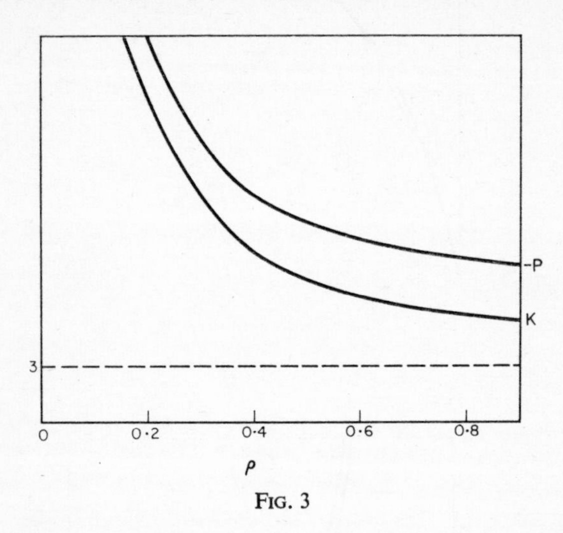

FIG. 3

in such a way that the lowest energy value be again -3. The lower line in Fig. 2 gives the relation obtained in this way between $v_0 = a/1{\cdot}4$ and $\rho = 0{\cdot}63/b$. It runs very near to the line obtained for the potential-function (1). A more exact calculation would show that it runs yet a little lower than shown in Fig. 2.

It is clear now what the cause of the large mass defect of He may be. The total energy of the He consists of four potential energies (the attraction of both protons on both neutrons) and only four kinetic energies, as contrasted to one potential energy and two kinetic energies in the H^2. In He therefore, the former will overcompensate the latter much more than in H^2. A similar phenomenon exists also in atomic spectra: the lowest energy value of the He is four times larger than that of H, because the ratio of the terms of potential to kinetic energy is 2 : 1 instead of 1 : 1 in H. The conditions are still more pronounced in the nucleus.

3

Before making the actual calculation for He, a remark on the existence of H^3 should be added. The Schrödinger equation $H\psi = E\psi$ for two neutrons 1, 2 and a proton 3 is

$$-10\left(\frac{\partial^2}{\partial r_{23}^2}+\frac{2}{r_{23}}\frac{\partial}{\partial r_{23}}+\frac{\partial^2}{\partial r_{13}^2}+\frac{2}{r_{13}}\frac{\partial}{\partial r_{13}}+\frac{\partial^2}{\partial r_{12}^2}+\frac{2}{r_{12}}\frac{\partial}{\partial r_{12}}+\right.$$
$$+\cos(213)\frac{\partial^2}{\partial r_{12}\partial r_{13}}+\cos(123)\frac{\partial^2}{\partial r_{12}\partial r_{23}}+$$
$$\left.+\cos(132)\frac{\partial^2}{\partial r_{13}\partial r_{23}}\right)\psi+(V(r_{13})+V(r_{23}))\psi$$
$$= E\psi(r_{23}, r_{13}, r_{12}) \qquad (5)$$

where (213) is the angle with the vertex 1 and the sides through 2 and 3. Assuming that $\psi(r_{13})$ is the solution of the Schrödinger equation (2) between the neutron 1 and the proton 3, it is reasonable to try the wave function

$$\psi_0 = \psi(r_{13})\psi(r_{23}) \qquad (6)$$

for (5). Actually, by calculating the expectation value for the energy $E_0 = (\psi_0, H\psi_0)$ of ψ_0 we obtain -2ε. Therefore the binding energy of the second neutron is certainly even larger than that of the first.† This is independent of the potential function. The conditions will remain similar if we complete the odd number of protons or neutrons to an even number.

In order to have a better value for the mass defect of H^3 than -2ε the Hassé variational method‡ may be tried. We calculate

$$(H-E_0)\psi_0 = -10\cos(132)\psi'(r_{13})\psi'(r_{23}) \qquad (7)$$

† This is, of course, not true for the third neutron as a wave function like (6) is not allowable for more than two neutrons in consequence of Pauli's principle. Actually if ρ is not too large, the third neutron has no positive binding energy.
‡ H. R. Hassé, *Proc. Camb. phil. Soc.* **26**, 542 (1930).

and choose α in

$$\psi_1 = \psi_0 + \alpha(H - E_0)\psi_0 \tag{8}$$

in such a way that $(\psi_1, H\psi_1)$ assumes its minimal value

$$E_1 = \tfrac{1}{2}E_0 + V'/2V_2 - [\tfrac{1}{4}(V'/V_2 - E_0)^2 + V_2]^{\frac{1}{2}} \tag{9}$$

where $E_0 = (\psi_0, H\psi_0) = -2\varepsilon$, $V_2 = (\psi_0, (H-E_0)^2\psi_0)$ and $V' = (\psi_0, (H-E_0)^2 H\psi_0)$. In the present case we have $V_2 = (1/3)K^2$ where K is the mean kinetic energy in H^2. This is very large and shows that ψ_0 is certainly rather far from the correct wave function. As to E_1, however, as V' is even larger than $(V_2)^{\frac{3}{2}}$ it turns out that it is not far from the value $E_0 = -2\varepsilon$. It might be therefore that the second neutron is only somewhat (perhaps twice) as strongly bound as the first. The relative occurrence of H^3 would be therefore much rarer even that that of H^2, as usually an isotope with a mass number one larger than the other is very rare if the mass defect is so small. The magnitude of the mass defect and even the existence of the H^3 becomes uncertain of course, if we assume repulsive forces between the neutrons.

4

We now come to the calculation of the binding energy of the He nucleus, 1 and 2 are neutrons, 3 and 4 protons. We may try as the first approximation to ψ the following expression

$$\psi_1 = \frac{f(r_{13})f(r_{23})f(r_{14})f(r_{24})}{[\int f(r_{13})^2 f(r_{23})^2 d3]^{\frac{1}{2}}} \tag{10}$$

where $\int \ldots d3$ indicates integration over all coordinates of the particle 3 and f is an as yet unknown function which will later be taken as the solution of an equation similar to (2) but with a different V. The meaning of (10) is, that the probability $\int \psi_1{}^2 d3$ of a certain position of 4 for a given position of 1 and 2 is $f(r_{14})^2 f(r_{24})^2$, in analogy to (6). Really, ψ will be symmetric with respect to the interchange of the pair 1, 2 with the pair, 3, 4.

Upon calculating the expectation value of the potential energy for ψ_1, given by (10), one obtains $4\int f(r_{12})^2 V(r_{12}) d1 d2 = 4P_f$, four times the mean potential energy of a nucleus H^2 in the state $\psi = f$.

For the kinetic energy, however, one gets

$$4K_f - 10\int[\int \cos(314) f'(r_{14}) f(r_{14}) f'(r_{13}) f(r_{13}) f(r_{23})^2 \times \times f(r_{24})^2 d3d4 / \int f(r_{13})^2 f(r_{23})^2 d3] d1d2 \qquad (11)$$

where

$$K_f = 10\int f'(r_{23})^2 d1d3 \qquad (12)$$

is the mean kinetic energy of the proton and neutron in H^2 in the state $\psi = f$. The kinetic energy for (10) is, in consequence of (11), smaller than $4K_f$ as the integral in (11) is positive. One sees this by writing

$$\cos(314) = \cos(312)\cos(214) + \sin(312)\sin(314)\cos\alpha \qquad (13)$$

where α is the angle between the planes through 1, 3, 2 and through 1, 3, 4. After inserting (13) into (11) one sees that the integral arising from the second part of (13) will vanish upon integration over α and the total integral in (11) becomes

$$\int \frac{(\int \cos(312) f'(r_{13}) f(r_{13}) f(r_{23})^2 d3)^2}{\int f(r_{13})^2 f(r_{23})^2 d3} d1d2 \qquad (14)$$

which is clearly positive. This integral was estimated in the following way. The function $f(r)$ was approximated by $\alpha \exp(-\beta r^2)$ with undetermined α and β. Then (14) was calculated and compared with the integral occurring in the expression (12) of the kinetic energy. It was found—as might be expected—that the ratio of both is independent from α and β and is equal to 0·5. For $f = \alpha \exp(-\beta r)$ the ratio is even 0·64, but 0·5 was adopted in the subsequent calculation in order to stay on the safer side. The total energy therefore is for ψ_1

$$(\psi_1, H\psi_1) = 4P_f + 3{\cdot}5K_f = 3{\cdot}5(K_f + 1{\cdot}14P_f) \qquad (15)$$

Now we can choose f so as to minimize (15), which is readily obtained, by assuming that f is the solution of a differential equation like (2), only with the potential multiplied by 1·14. The total energy (15) is then 3·5 times the binding energy of such an imaginary nucleus, with 1·14 times the real attraction.

As was pointed out before, the characteristic value of (2) is under these conditions very sensitive to a small increase of the potential.

For $\rho = 0{\cdot}22$, $v_0 = 140$ we obtain in this way a mass defect for the He, which is 7·85 times larger than that of the H^2.

One more improvement was made in this calculation: instead of ψ_1 the symmetrized function

$$\psi = \psi_1 + \psi_2 \tag{16}$$

with

$$\psi_2 = \frac{f(r_{13})f(r_{23})f(r_{14})f(r_{24})}{[\int f(r_{13})^2 f(r_{14})^2 d1]^{\frac{1}{2}}} \tag{10a}$$

was taken as wave function. Then the expectation value for the energy becomes

$$\frac{(\psi_1, H\psi_1) + (\psi_1, H\psi_2)}{1 + (\psi_1, \psi_2)} = \frac{A+B}{1+S} \tag{17}$$

A is, according to (15) $4P_f + 3{\cdot}5K_f$. In order to calculate S and B, the function $f(r)$ was again approximated by $\alpha \exp(-\beta r^2)$, which gave $S = 0{\cdot}84$ and $B = 3{\cdot}82P_f + 2{\cdot}80K_f$ so that in the whole we have

$$E = 3{\cdot}45(K_f + 1{\cdot}24P_f) \tag{18}$$

Minimizing (18) in the same way we get 12 for the ratio of the mass defect of He and H^2 in contrast with the observed value of about 17. This again corresponds to $\rho = 0{\cdot}22$ or a half-width of about $0{\cdot}38e^2/mc^2$ for the potential hole. For larger ρ the ratio becomes smaller, for smaller ρ larger. Another possibility is to take $\psi = (\psi_1\psi_2)^{\frac{1}{2}}$. This gives on a similar calculation $E = 3{\cdot}2(K_f + 1{\cdot}25P_f)$ for $\rho = 0{\cdot}22$ or a ratio of $11\frac{1}{2}$. Now one could take a linear combination of this ψ with that of (16), which would give a still somewhat lower value—not very much, however, because the two wave functions forming the linear combination do not differ very much from each other.†

There is, however, another possibility to improve the wave function, namely to take advantage of the mixed differential coefficient terms in (5) as we did it for H^3 with the Hassé method. This would for the ρ under consideration, probably increase the ratio even somewhat over the experimental value.

† The best wave function I could find was $\alpha \exp[-\frac{1}{2}\beta(r_{13}^2 + r_{23}^2 + r_{14}^2 + r_{24}^2 + \frac{1}{2}r_{12}^2 + \frac{1}{2}r_{34}^2)]$, it gave a ratio of about 14.

It seems therefore that if the potential hole is thin, the attractive forces between the neutron and proton give even a too large mass defect for He, so that a repulsion between the different neutrons and between the different protons may be assumed.

In conclusion one can state that if the basis of the present calculation should prove to be correct, the difference of the mass defects of He and H^2 can be attributed to the great sensitivity of the total energy to a virtual increase of the potential—as is brought about by the fact that every particle in the He is under the influence of two attracting particles, instead of one as in the case of H^2. The reason for this sensitivity lies in the functional dependence of the lowest energy value on a multiplicative factor v of the potential, which is as follows. For very small values of v there is no negative energy value at all (provided that the potential goes more strongly to zero than $1/r^2$). If v attains a critical value ($5/\rho^2$ for the potential (1)), there arises one discrete energy value at zero, which becomes more negative on a further increase of v. In the neighbourhood of the critical value, however, a very large relative change corresponds to a comparatively small relative change of v. A characteristic property in the neighbourhood of the critical value of v is that the mean kinetic energy is almost oppositely equal to the mean potential energy, i.e., the total negative energy is much smaller than the kinetic. That this is actually the case can be simply shown by an application of Heisenberg's indetermination principle.†

No similar sensitivity exists, of course, in one dimension as the critical value of v is 0 in this case.‡

† W. Heisenberg, *Z. Phys.* **77**, 1, calculated in this way the kinetic energy of the electron in the neutron and inferred from the number obtained that it cannot obey the laws of quantum mechanics since the mean kinetic energy is much larger than the mass defect. This consideration, of course, cannot be applied for the free electrons in the higher nuclei, *Z. Phys.* **50**, 548, as the mass defects are much larger than that used by Heisenberg for the neutron, and the same holds also for the nuclear diameters.

‡ R. Peierls, *Z. Phys.* **58**, 59 (1929).

7

On the Scattering of Neutrons by Protons†

E. WIGNER

ABSTRACT

In calculating the scattering of neutrons by protons we assume that the interaction is given by a short-range potential. The relation between this potential and the mass-defect of Urey's hydrogen (deuterium) will be investigated.

1

THE scattering of neutrons by nuclei has been treated by Massey.‡ If we investigate this problem again, it is because Massey was concerned mainly with the scattering by heavy nuclei while we shall treat the scattering by protons and its relation with the mass-defect of H^2. We shall also discuss the angular dependence of the scattering intensity.

It is possible to calculate the scattering by Faxén's and Holtsmark's method:§ The incoming monochromatic plane wave is first split into waves whose total angular momentum has definite values, and one works with a centre-of-mass-coordinate system; thus, the wave function depends only on the differences x, y, z of the coordinates of protons and neutrons respectively. Let the incoming wave be $\exp(ipz/\hbar)$ and the wave function of a state with total angular momentum l and z-component equal to zero be ψ_l. Thus we obtain

† *Zeits. f. Phys.* **83,** 253 (1933).

‡ H. S. W. Massey, *Proc. Roy. Soc.* (*London*), **A138,** 460 (1932).

§ H. Faxén and J. Holtzmark, *Z. Phys.*, **45,** 307 (1927).

$$\exp(ipz/\hbar) = a_0\psi_0 + a_1\psi_1 + a_2\psi_2 \tag{1}$$

where

$$\psi_0 = r^{-1}\sin pr/\hbar; \quad \psi_1 = \partial\psi_0/\partial z$$

$$\psi_2 = 2\left(\frac{\partial^2}{\partial z^2} - \frac{\partial^2}{\partial x^2} - \frac{\partial^2}{\partial y^2}\right)\psi_0$$

etc.

Comparing the various differential coefficients with respect to z on both sides of equation (1) at the origin we obtain the coefficients most easily. The first $l-1$ differential coefficients of ψ_l vanish at this point and we obtain

$$\exp(ipz/\hbar) = \frac{\sin pr/\hbar}{pr/\hbar} - \frac{3i\hbar}{p}\cos\theta\,\frac{\partial}{\partial r}\,\frac{\sin pr/\hbar}{pr/\hbar} + \ldots \tag{1a}$$

where θ is the angle between the z-axis and the direction x, y, z. The various terms in equation (1) are solutions of the free-particle Schrödinger equation $-\hbar^2\nabla^2\psi_l = p^2\psi_l$ which have to be replaced by solutions φ_l of the true Schrödinger equation ($\hbar$ is Planck's constant divided by 2π)

$$-\hbar^2\nabla^2\varphi_l = (p^2 - MV(r))\varphi_l \tag{2}$$

where M is the proton mass and p the momentum of the particles in the centre-of-mass system, so that the relative velocity is $2p/M$. We substitute ψ_l by φ_l by adding functions f_l to ψ_l which have the same angular momentum as the corresponding ψ_l and which behave like an outgoing wave at infinity, i.e. at infinity they are proportional to the following expressions

$$r^{-1}\exp(ipr/\hbar), \qquad \frac{\partial}{\partial z}r^{-1}\exp(ipr/\hbar)$$

$$\left(2\frac{\partial^2}{\partial z^2} - \frac{\partial^2}{\partial x^2} - \frac{\partial^2}{\partial y^2}\right)r^{-1}\exp(ipr/\hbar)$$

The scattered wave then is $f_0 + f_1 + f_2 + \ldots$

The scattering experiments made by I. Curie and Joliot† show that the scattering is essentially spherically symmetric and that the scat-

† I. Curie et F. Joliot, *La projection des noyaux atomiques par un rayonnement tres pénétrant*. Paris (1932). J. L. Destouches, *Etat actuel de la théorie du neutron*. Paris (1932).

tering in the forward direction is not as predominant as with α-particles. More recent investigations by Dunning and Pegram† confirm this general feature of scattering. Therefore it seems obvious to assume that the scattered wave consists basically of the spherically symmetric component f_0 only. This would agree with the fact that $\psi_1, \psi_2 \ldots$ are almost solutions of Schrödinger equations, i.e. the term $MV\varphi_l$ in equation (2) is small for $l \neq 0$. This certainly holds if the distance a beyond which V is very small is considerably shorter than the wavelength $\hbar/p$. Thus we assume that the distance at which the potential is half its maximum value is smaller than e^2/mc^2.‡

2

We now carry out the calculation, first with a square-well potential which is $-v$ for $r < a$, 0 for $r > a$. The wave function of the *discrete* stationary state is $r^{-1} \sin p_0 r/h$ for $r < a$, $cr^{-1} \exp [ip_i(r-a)/\hbar]$ for $\mathbf{r} > a$, where p_i is positive imaginary, $-p_i^2/M = \varepsilon$ the binding energy [according to Bainbridge§ about three times the rest energy of the electron] and $p_0^2 = M(v-\varepsilon)$. Because of the continuity of the wave-function and its differential coefficients we obtain, as is well known, $(a/\hbar = a')$

$$p_0 \cot p_0 a' = ip_i = -(M\varepsilon)^{\frac{1}{2}} \tag{3}$$

that is

$$(M(v-\varepsilon))^{\frac{1}{2}} a' > \frac{\pi}{2}$$

in order that a stationary state exist. By series expansion (putting $p_0 a' = \pi/2 + x$ and expanding in powers of x) we obtain

$$(M(v-\varepsilon))^{\frac{1}{2}} = \frac{\pi}{2a'} + \frac{2(M\varepsilon)^{\frac{1}{2}}}{\pi} - \frac{8M\varepsilon a'}{\pi^3} \tag{3a}$$

The smaller a' is, the better this series converges. For all really important a' the first two or three terms are sufficient.

Now we calculate f_0, i.e. the spherically symmetric part of the

† J. R. Dunning and G. B. Pegram, *Phys. Rev.* **43**, 497 (1933).
‡ Cf. also E. Wigner, *Phys. Rev.* **43**, 252 (1933) (see p. 170 of this book).
§ K. T. Bainbridge, *Phys. Rev.* **42**, 1 (1932); J. D. Hardy, E. F. Barker and D. U. Dennison, *Phys. Rev.* **42**, 279 (1932).

scattered wave. For $r < a$ we have again $\varphi_0 = cr^{-1} \sin p_i r/\hbar$, for $r > a$, however,

$$\varphi_0 = \frac{\sin p_a r/\hbar}{p_a r/\hbar} + br^{-1} \exp\,[ip_a(r-a)/\hbar] \tag{4}$$

where p_a is real this time. From this follows

$$b = -\frac{\hbar}{p_a} \frac{p_i \cot p_i a' \sin p_a a' - p_a \cos p_a a'}{p_i \cot p_i a' - ip_a} \tag{4a}$$

or, by putting $p_i^2 = p_a^2 + Mv$ and using the series we obtain equation (4b)

$$b = -\hbar/(ip_a + (M\varepsilon)^{\frac{1}{2}}) - \hbar a^1/2 + \;.\;.\;. \tag{4b}$$

The first approximation of Born's scattering theory would yield for b†

$$b = \frac{\hbar Mv}{2p_a^2} \exp\,(ip_a a')\left(a' - \frac{1}{2p_a} \sin 2p_a a'\right)$$

and would not be sufficient for our case.

We now calculate f_1. For $r < a$ we have

$$\varphi_1 = c \cos\theta \frac{\partial}{\partial r} \frac{\sin p_i r/\hbar}{p_i r/\hbar}$$

for $r > a$, however,

$$\varphi_1 = -\frac{3i\hbar}{p_a} \cos\theta \frac{\partial}{\partial r} \frac{\sin p_a r/\hbar}{p_a r/\hbar} + b \cos\theta \frac{\partial}{\partial r} \frac{\exp\,[ip_a(r-a)/\hbar]}{ip_a r/\hbar} \tag{5}$$

The continuity condition for $r = a$ yields

$$b = \left(\frac{12}{\pi^2} - 1\right) i\hbar p_a a'^2 \tag{5a}$$

Thus the scattering of the neutrons with angular momentum 1 is of an order of magnitude we neglected in (4b), but we retain equation (5) so that we can judge the importance of the scattering in the forward direction.

The total scattered wave for $r \to \infty$ is

$$f_0 - f_1 + \ldots = r^{-1} \exp\,[ip_a(r-a)/\hbar] \times$$
$$\times \left(-\frac{\hbar}{(M\varepsilon)^{\frac{1}{2}} + ip_a} - \frac{a}{2} + \frac{0{\cdot}21 ip_a a^2}{\hbar} \cos\theta\right) \tag{6}$$

† Cf. L. Brioullin, *Die Quantenstatistik*, p. 262, Berlin (1931).

The intensity of the scattering in the centre-of-mass system in the θ-direction is

$$J_\theta = \left(\frac{\hbar(M\varepsilon)^{\frac{1}{2}}}{M\varepsilon+p_a^2}+\frac{a}{2}\right)^2+\left(\frac{p_a\hbar}{M\varepsilon+p_a^2}+\frac{0{\cdot}21 p_a a^2}{\hbar}\cos\theta\right)^2 \tag{6a}$$

Considerable scattering in the forward direction can be expected if p_a is not much smaller than $(M\varepsilon)^{\frac{1}{2}}$ (a condition most experiments satisfy) and if $0{\cdot}21\ a^2(p_a^2+M\varepsilon)/\hbar^2$ is not much smaller than 1. The asymmetry of the scattering in the centre-of-mass system is a measure for the range of the interaction-potential between neutrons and protons. Favoured scattering in the backward direction can only be expected, if this potential has repulsive regions also.

In order to calculate the interaction cross section q of protons and neutrons we must remember that the density of both types of particles is 1 for the wavefunction equation (1) and that the number of collisions per cm^3/sec is $2\,qp_a/M$ (as the relative velocity is $2\,p_a/M$). The number of the scattered pairs with their centre of mass in the investigated cm^3 and their separation somewhere between r and $r+1$ is thus exactly q. By squaring equation (6) and integrating in this region we obtain for this term (excluding terms of order a^2)

$$q = \frac{8\pi\hbar^2}{M}\,\frac{1+a(M\varepsilon)^{\frac{1}{2}}/\hbar}{E+2\varepsilon} \tag{7}$$

where E is the kinetic energy $2\,p_a^2/M$ of the neutron with the proton at rest.

With Eckart's potential† $V(r) = -4v_0/(1+\exp(r/\rho)(1+\exp(-r/\rho)$ the expression $1+4\rho(M\varepsilon)^{\frac{1}{2}}/\hbar$ replaces the factor $1+a(M\varepsilon)^{\frac{1}{2}}/\hbar$. Eckart's potential follows most closely the square-well potential with $a = 2{\cdot}5$ and $v = 0{\cdot}61\ v_0$. The two expressions for the scattering are not very different.

At energies of $0{\cdot}5\times10^6$ to 2×10^6 eV measurements by Meitner and Philipp‡ found an interaction radius for neutrons which was bigger than 8×10^{-13} cm. If $a = 0$ equation (7) yields for these energies values of 8×10^{-13} and 10×10^{-13} cm respectively; for

† C. Eckart, *Phys. Rev.* **35,** 1303 (1930).
‡ L. Meitner and K. Philipp, *Naturwissenschaften*, **20,** 929 (1932).

bigger a they are bigger still. Values calculated from measurements by I. Curie and Joliot† are much smaller and contradict equation (7). The measurements by Dunning and Pegram, however, agree with the ones by Meitner and Philipp.

3

In the simplest case we assume a to be equal to zero, i.e. we assume an infinitely large potential with an infinitely small range. The scattering cross section will then be independent of the potential. For this reason and because Meitner and Philipp's experiments do not contradict this assumption for the time being we shall discuss its implications in more detail.

The wave-function satisfies equation $=\hbar^2\nabla^2\psi = i\hbar M\partial\psi/\partial t$ for every non-zero distance between the particles and the potential acts only through a singular behaviour of the wave-function for $r = 0$‡. If we expand the wave-function in spherical functions of the direction θ, φ of the line joining the particles we obtain

$$\psi = \sum_{l=0}^{\infty} \sum_{m=-l}^{l} \varphi_{lm}(\xi, \eta, \zeta, r) P_{lm}(\theta, \varphi) \tag{8}$$

where ξ, η, ζ are the centre-of-mass coordinates and r is the distance between the particles. Thus ϕ_{lm} with $l \neq 0$ will be regular also for $r = 0$ or even vanish. The smaller the radius of the potential the less will it affect the φ_{lm} with $l \neq 0$. This follows for ϕ_{10} also from equation (5): b vanishes if a tends to zero.

In φ_0, however, we have a singularity if $r = 0$. In equation (4) we calculated the stationary states. They are if $r > a$ always

$$\varphi_0 = \frac{\sin pr/\hbar}{pr/\hbar} + \frac{b}{r} \exp(ipr/\hbar) = c\left(\frac{\sin pr/\hbar}{pr/\hbar} + b' \frac{\cos pr/\hbar}{r}\right) \tag{9}$$

† *Loc. cit.*

‡ The possibility of certain singularities in the wave-function and in particular the necessity to assume such singularities with certain singular potentials ($1/r^n$ if $n > 2$) was investigated by J. v. Neumann in a hitherto unpublished paper. I would like to thank Mr. Neumann here very much for allowing me to study his unpublished paper and for many valuable suggestions.

where b' is obtained from equation (4b) and is

$$b' = \frac{b}{1+ibp/\hbar} = -\frac{\hbar}{(M\varepsilon)^{\frac{1}{2}}} \tag{9a}$$

so that the coefficients of the term 1/r and of the constant term are independent of the energy parameter. Since every wave-function can be written as a linear combination of stationary states, this holds for all wave-functions. We may leave out the potential completely and replace it by this limiting condition where $-\varepsilon$ is the only discreet eigenvalue.

Only further experiments will show if it is possible to use this assumption of the potential for describing scattering experiments and even, whether a potential can be used at all. As is known, this potential cannot be generalized further for Dirac's electron.†

† Cf. J. L. Destouches, *loc. cit.*

8
An Electrical Quadrupole Moment of the Deuteron†‡

J. M. B. KELLOGG, I. I. RABI, N. F. RAMSEY, JR.
AND J. R. ZACHARIAS

THE molecular beam magnetic resonance method§ applied to HD molecules at liquid nitrogen temperature gives the magnetic moment of the proton and the deuteron. When applied to H_2 and D_2 molecules the method reveals the close groups of sharp resonance minima shown in Figs. 1 and 2. The resonance minima for H_2 agree in number and location with predictions based on the assumption of spin–spin magnetic interaction of the two protons

$$(\boldsymbol{\mu}_1 . \boldsymbol{\mu}_2)/r^3 - 3(\boldsymbol{\mu}_1 . \boldsymbol{r})(\boldsymbol{\mu}_2 . \boldsymbol{r})/r^5)$$

and a spin–orbit interaction of the protons with the rotation of the molecule $(2\mu_P \overline{H} I . J)$. All symbols here have their usual significance except $\overline{H}$ which is the spin–orbit interaction constant. The only state of the molecule to be considered is the lowest rotational state of orthohydrogen: $J = 1$ and total nuclear spin $I = 1$. The nine energy levels which arise from these interactions and from the external magnetic field give six possible transitions for the nuclear spin because $\Delta M_N = \pm 1$. The pattern can be accounted for completely on

† *Phys. Rev.* **55**, 318 (1939).
‡ Publication assisted by the Ernest Kempton Adams Fund for Physical Research of Columbia University.
§ J. M. B. Kellogg, I. I. Rabi, N. F. Ramsey and J. R. Zacharias, *Bull. Amer. Phys. Soc.* Vol. 13, No. 7, Abs. 24 and 25.

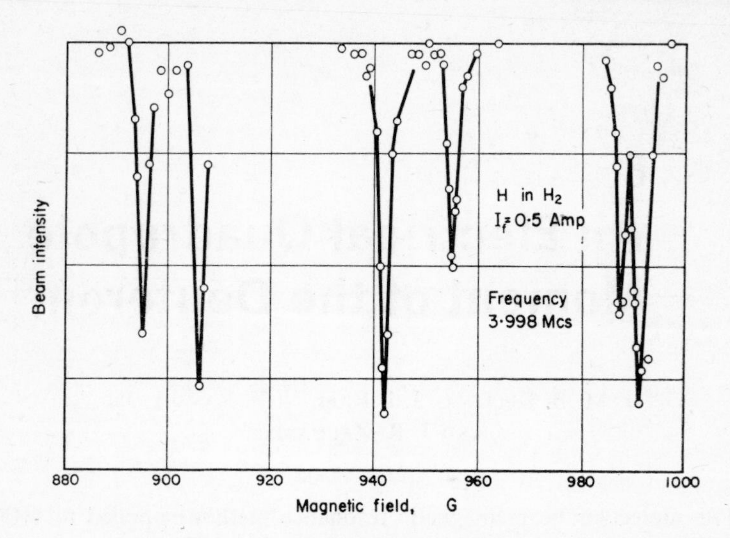

FIG. 1. Resonance minima for H in H_2.

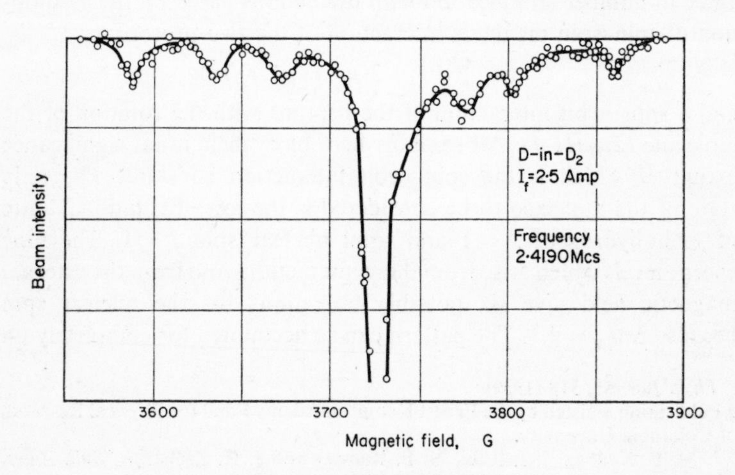

FIG. 2. Resonance minima for D in D_2.

the basis of the known value of the proton moment ($\mu_P = 2{\cdot}78$) with the known value of the internuclear distance and a value for $\overline{H}$ of 27 gauss.

In the case of D_2 the deep central minimum arises from the states with $I = 2$ and $J = 0$. The six smaller peaks arise from the states with $J = 1$ and $I = 1$. The states with $J = 2$ are not abundant enough at these low temperatures for observation. Since the internuclear distance in the D_2 molecule is the same as in H_2 and the mass is twice as great the spin–orbit interaction constant $\overline{H}$ is half as great. The deuteron magnetic moment ($\mu_D = 0{\cdot}853$) is 0·307 times that of the proton and therefore the magnetic spin–spin interaction is proportionately smaller. It was expected that the theory of the resonance minima for H_2 when applied to D_2 should give the locations of the minima from the constants given above. The displacements of the minima from the centre should be much less than those of H_2. However experiment shows the displacements to be six times greater than the predicted values.

This effect can be accounted for by the presence of an electrical quadrupole moment in the deuteron. The interaction energy which gives rise to the large displacements is that of the nuclear quadrupole moment with the gradient of the molecular electric field. This form of interaction contributes to the nine energy levels exactly as the spin–spin and therefore appears as a larger spin–spin interaction.

To prove that the large displacements in D_2 are of nuclear origin rather than molecular, similar experiments were performed on the proton and the deuteron in the HD molecule. The group of resonance minima for H was narrow as expected and that for D had large displacements as in D_2 Furthermore, the experimentally evaluated spin–orbit interaction constant for D_2 is one-half as great as that for H_2 as predicted. We therefore believe that the apparent large spin–spin interaction is not magnetic, nor is it of molecular origin and must be a nuclear effect which behaves like a quadrupole moment.

To obtain the magnitude of this quadrupole moment one must know the molecular electric field. This value can be calculated from the various wave functions which have been suggested for the hydrogen molecule. The result of such a calculation by Dr. A.

Nordsieck with Wang wave functions when combined with our data yields a quadrupole moment $Q = (3z^2 - r^2)_{Av}$ of about 2×10^{-27} cm^2. The chief source of error lies in the inaccuracy of the wave functions.

The sign of the quadrupole moment may also be inferred from our measurements in two ways. Present indications are that it is positive, that is, the charge configuration is prolate along the spin axis. Full details of these experiments will be published later in this journal.

We are greatly indebted to Dr. Nordsieck for making available to us the results of his calculations and to Dr. Brickwedde of the National Bureau of Standards for preparation of the sample of pure HD used in the experiments. The experiments were supported in part by a grant from the Research Corporation.

9

On Nuclear Forces†

B. CASSEN AND E. U. CONDON

The various types of exchange forces that are being used in current discussions of nuclear structure may all be simply expressed in terms of a formalism which attributes five coordinates to each "heavy" particle and applies the Pauli exclusion principle to all the particles in the system. The simplest assumption for the interaction law is that which implies equality of proton–proton and neutron–neutron forces and also equality with the proton–neutron forces of corresponding symmetry. This is in accord with the empirical knowledge of these interactions at present.

IN this paper we show how the use of a coordinate having two proper values which tells whether a particle is a proton or a neutron, together with the assumption of the Pauli exclusion principle for all the particles, gives a unified description of the various types of exchange forces used in nuclear structure theories. Such a coordinate was first introduced by Heisenberg‡ and also plays a role in the Fermi–Konopinski–Uhlenbeck§ theory of beta disintegration.

We suppose that each heavy particle (proton or neutron) is described by five coordinates. These are three for its position in space, a spin coordinate σ giving the component of its angular

† *Phys. Rev.* **50,** 846 (1936).
‡ Heisenberg, Z. *Phys.* **77,** 1 (1932).
§ Fermi, Z. *Phys.* **88,** 161 (1934); Konopinski and Uhlenbeck, *Phys. Rev.* **48,** 7 (1935).

momentum along some direction in space, and a fifth coordinate, τ, which can have the values ± 1. If τ has the value $+1$ the particle is a proton, while the value -1 indicates that it is a neutron.

The spin angular momentum is a vector equal to $\frac{1}{2}\hbar$ times the vector, $\boldsymbol{\sigma}$, which is represented by

$$\boldsymbol{\sigma} = \begin{pmatrix} \mathbf{k} & \mathbf{i}-i\mathbf{j} \\ \mathbf{i}+i\mathbf{j} & -\mathbf{k} \end{pmatrix}$$

the rows and columns referring to states which are labelled by precise values of the z component of $\boldsymbol{\sigma}$. This nonrelativistic description of the spin was introduced by Pauli and by Darwin.

In the same way τ can be considered purely formally like the z component of a vector. The analogy is purely formal in that the three "components" of $\boldsymbol{\tau}$ do not refer to directions in space. Formally we may write

$$\boldsymbol{\tau} = \begin{pmatrix} \mathbf{n} & \mathbf{l}-i\mathbf{m} \\ \mathbf{l}+i\mathbf{m} & -\mathbf{n} \end{pmatrix}$$

where $\mathbf{l}$, $\mathbf{m}$ and $\mathbf{n}$ behave algebraically like the three unit vectors $\mathbf{i}$, $\mathbf{j}$ and $\mathbf{k}$. The third component of $\boldsymbol{\tau}$ may be called the character coordinate and the whole expression $\boldsymbol{\tau}$ the character vector.

We postulate that in an assembly of heavy particles the wave function has to be antisymmetric in all particles with regard to exchange of all five of their coordinates. We want to show that this gives a convenient formalism for working with nuclear problems.

Let us first consider any attribute of a single heavy particle such as its mass, its charge or its magnetic moment. If A is the arithmetic mean of the two values for proton and neutron and B is half the difference, proton value minus neutron value, then that attribute will appear in the equations as a term involving,

$$(A+B\tau)$$

For example, the electrostatic charge will appear as $\frac{1}{2}e(1+\tau)$ where e is the electronic charge.

Next, let us consider the scalar product $\boldsymbol{\tau}_1 . \boldsymbol{\tau}_2$ of the character vectors associated with two particles. We have

$$(\boldsymbol{\tau}_1+\boldsymbol{\tau}_2)^2 = \boldsymbol{\tau}_1^2+\boldsymbol{\tau}_2^2+2\boldsymbol{\tau}_1 . \boldsymbol{\tau}_2$$

since the operators for two different particles commute. Now as defined τ is formally like twice an angular momentum vector of magnitude $\frac{1}{2}$. Therefore, the possible values of the vector sum are twice 1 and zero. Letting $2T$ stand for the magnitude of the resultant we have

$$4T(T+1) = 3+3+2\tau_1 . \tau_2$$

so the allowed values of $\tau_1 . \tau_2$ are $+1$ and -3. The value $+1$ corresponds to the case of parallel character vectors and so to a wave function that is symmetric in τ_1 and τ_2 while the value -3 corresponds to resultant zero of the two character vectors and hence to an antisymmetric dependence of the wave function on τ_1 and τ_2.

Therefore, the expression

$$\tfrac{1}{2}(1+\tau_1 . \tau_2)$$

has the allowed values $+1$ and -1, the positive value going with wave functions symmetric in τ_1 and τ_2, while the negative value has for its proper function a wave function antisymmetric in these two character coordinates.

These results are, of course, exactly analogous to the well-known results for the vector sum of two spin angular momenta and their connection with the symmetry properties of the wave function with regard to exchange of σ_1 and σ_2.

The applicability of the Pauli exclusion principle to a dynamical system requires that the Hamiltonian function for the system be a symmetric function of the coordinates of the particles. In looking for possible interaction laws we therefore have to confine ourselves to symmetric functions.

So far, four types of exchange forces have been proposed for description of the interaction between heavy particles. These are:

1. Ordinary (Wigner) potential.† This is the familiar kind and is simply a function of the distance between the two particles.

2. Heisenberg potential.‡ This is of the form of a function of the distance multiplied by an operator H. This operator is defined as having the value $+1$ when applied to a wave function that is sym-

† Wigner, *Phys. Rev.* **43,** 252 (1933) (see p. 170 of this book).
‡ Heisenberg, *Z. Phys.* **77,** 1 (1932) (see p. 144 of this book).

metric with regard to exchange of both position and spin coordinates of the two particles whose interaction is being considered, and the value -1 for the antisymmetric case.

3. Bartlett potential.† This is a function of the distance multiplied by an operator B. This operator is defined as having the value $+1$ when applied to a wave function that is symmetric in the spin coordinates alone and -1 for the antisymmetric case.

4. Majorana potential.‡ This is a function of the distance multiplied by an operator M. This operator is defined as having the value $+1$ when applied to a wave function that is symmetric with regard to exchange of the positional coordinates only of the two particles in question, and -1 when applied to an antisymmetric function in the positional coordinates.

Evidently the Majorana type can be expressed in terms of the preceding two:

$$M = HB = BH$$

Since the operator H exchange both position and spin, and the Bartlett operator B exchanges spin only, the product will be equivalent to an exchange of position only, for the double exchange of spin provided by the combined action of H and B cancels out and is the same as no exchange of spin.

We now point out that the four operators, 1, H, B and M are readily expressible in terms of the spin and character vectors, $\boldsymbol{\sigma}$ and $\boldsymbol{\tau}$ of the two particles. This follows from the requirement of over-all antisymmetry of the wave functions in the five coordinates of each of the particles. Let the letters $a, b, c, d\ldots$ stand for the different particles and consider a general wave function ψ that is a function of all five of the coordinates of each particle. More explicitly

$$\psi = \psi(\mathbf{r}_a, \tau_a, \sigma_a; \mathbf{r}_b, \tau_b, \sigma_b; \mathbf{r}_c, \tau_c, \sigma_c; \ldots)$$

Whatever the functional form of ψ this can be written

$$\psi = \tfrac{1}{2}[\psi(\mathbf{r}_a, \tau_a, \sigma_a; \mathbf{r}_b, \tau_b, \sigma_b; \ldots) + \psi(\mathbf{r}_b, \tau_a, \sigma_b; \mathbf{r}_a, \tau_b, \sigma_a; \ldots)] +$$
$$+ \tfrac{1}{2}[\psi(\mathbf{r}_a, \tau_a, \sigma_a; \mathbf{r}_b, \tau_b, \sigma_b; \ldots) - \psi(\mathbf{r}_b, \tau_a, \sigma_b; \mathbf{r}_a, \tau_b, \sigma_a; \ldots)]$$

† Bartlett, *Phys. Rev.* **49**, 102 (1936).

‡ Majorana, *Z. Phys.* **82**, 137 (1933) (see p. 161 of this book).

that is, as the sum of a function symmetric in the position and spin coordinates of particles a and b and one antisymmetric in these same coordinates. As we require ψ to be antisymmetric in all five coordinates of a and b we know that the first term here must be antisymmetric in τ_a and τ_b and the second term must be symmetric in τ_a and τ_b. Therefore the operator H has the value -1 for symmetry in τ_a and τ_b, and $+1$ for antisymmetry in τ_a and τ_b. Using the earlier calculation of $\boldsymbol{\tau}_1 . \boldsymbol{\tau}_2$ we have

$$H_{ab} = -\tfrac{1}{2}(1+\boldsymbol{\tau}_a . \boldsymbol{\tau}_b)$$

which expresses the Heisenberg exchange operator in terms of the two character vectors.

Similarly it is easy to see that

$$B_{ab} = +\tfrac{1}{2}(1+\boldsymbol{\sigma}_a . \boldsymbol{\sigma}_b)$$

and therefore, in view of the relation, $M = HB$, we have

$$M_{ab} = -\tfrac{1}{4}(1+\boldsymbol{\sigma}_a . \boldsymbol{\sigma}_b)(1+\boldsymbol{\tau}_a . \boldsymbol{\tau}_b)$$

which completes the expression of each of the exchange operators in terms of symmetric functions of the coordinates of the two particles.

With the different types of exchange operators written in this simple way it suggests itself that the general law of interaction for the specifically nuclear forces can be written in the form:

$$U = V + V_h H + V_b B + V_m M$$

Here the four V's may be quite different functions of the separation distance but the simplest assumption is that the entire dependence of the interaction on $\boldsymbol{\sigma}$ and $\boldsymbol{\tau}$ is contained in the operators 1, H, B and M.

Of course, this simple result is not required by the formalism. It is simply the simplest form for the exchange operators. The mere requirement of a symmetric function would be met if any one or all of the V's were replaced by

$$A + B(\tau_1 + \tau_2) + C\tau_1\tau_2$$

where A, B and C are functions of the distance of separation. In fact, this more general form is necessary even for the description of the Coulomb interaction between the particles which for two particles is expressed as

$$\tfrac{1}{4}(e^2/r)(1+\tau_1)(1+\tau_2)$$

The expression above, involving A, B and C, has the value $(A+2B+C)$ for two protons, the value $(A-C)$ for a proton and a neutron, and the value $(A-2B+C)$ for two neutrons. If proton–proton forces are the same as neutron–neutron forces we may conclude that $B = 0$, and if like-particle forces are the same as proton–neutron forces in states of corresponding symmetry then we can conclude that $C = 0$. With both B and C equal to zero the dependence on the components τ_1 and τ_2 is gone and we are reduced to the simpler original form.

The assumption that B and C are zero seems to be in accord with the facts about nuclear interactions as far as these are known.† The assumption makes the unification that there are only four different force laws, corresponding to the four possible types of symmetry in σ and τ. These four types are describable in terms of more usual notation by giving the symmetry in position and spin, since this determines the symmetry in character. A state that is symmetric in spin is called a triplet, one antisymmetric a singlet. Symmetry for exchange of position will be denoted by S and antisymmetry by P, since these are the standard notations for states of least orbital angular momentum in the two-body problem which have these positional symmetry properties. Here, however, we use S and P in a more general sense.

The distinct laws of interaction are given in Table 1. Using the values of the operators 1, H, B and M we can write for the four interaction laws:

$$U(^3S) = V+V_h+V_b+V_m$$
$$U(^1S) = V-V_h-V_b+V_m$$
$$U(^3P) = V-V_h+V_b-V_m$$
$$U(^1P) = V+V_h-V_b-V_m$$

These are readily solved for explicit expressions for each V in terms of the four empirically occurring combinations.

† The consequences of assuming equality of the various specifically nuclear forces for like and unlike particles are considered in detail in a paper by Feenberg and Breit in *Phys. Rev.* **50,** 850 (1936) which we had the pleasure of seeing in manuscript after this paper was sent in.

TABLE 1

Symmetry in			Notation	Occurrence
Position	Spin	Character		
s	*s*	*a*	^3S	proton–neutron
s	*a*	*s*	^1S	proton–neutron proton–proton neutron–neutron
a	*s*	*s*	^3P	proton–neutron proton–proton neutron–neutron
a	*a*	*a*	^1P	proton–neutron

We shall only make a few brief remarks about the empirical facts as they are known as these have been recently reviewed by Bethe and Bacher.† Ideally one would like to learn all eight force laws (there are eight if we do not make the simple formal assumption of the previous section) from studies based wholly on the two-body problem. So far this is not possible.

The situation with regard to the two-body problems is this:

Proton–Neutron

$U(^3S)$: Normal state of deuteron. Observed binding energy gives a relation between depth and width of a potential well.

Scattering of neutrons by protons: This involves all four laws in principle, but in fact owing to short range of the forces only the two S laws enter in an important way for neutron energies less than some tens of millions of volts. Slow neutron scattering cross section indicates a ^1S level of deuteron near to zero binding energy, according to Wigner.

† Bethe and Bacher, *Rev. Mod. Phys.* **8,** 82 (1936).

Photodissociation of the deuteron: Electric dipole effect involves transitions from bound ^3S normal state to continuum of ^3P, hence these two laws. In addition to Bethe and Bacher, the problem is discussed by Breit and Condon.† Magnetic dipole effect produces transitions from ^3S normal state to ^1S continuum. This is important near the photoelectric threshold.

Radiative capture of neutrons by protons: Here the important effect is for slow neutrons principally by action of magnetic dipole radiation from ^1S continuum to ^3S normal state, according to Fermi.‡

None of these involve the ^1P law in an essential way. Apparently this can only be studied by scattering very high energy neutrons with protons.

Proton–Proton

Here a little evidence comes from the probable nonexistence of He^2. But mainly the knowledge comes from the recent work of Tuve, Heydenburg and Hafstad as analysed by Breit, Condon and Present. § The analysis indicates that up to 1 MeV the departures from coulomb scattering may be described entirely in terms of effects of the ^1S law and gives strong indication that this is the same as the ^1S law in the deuteron.

Neutron–Neutron

No positive evidence from two-body interactions. Absence of a double neutron is in accord with assumption of the same ^1S law as in proton–neutron since the ^1S level is now supposed to be virtual (see reference § for details).

All other knowledge of the force laws comes from approximate calculations of binding energies of many-body nuclei as fully reviewed by Bethe and Bacher. These are in accord with assumption of equality

† Breit and Condon, *Phys. Rev.* **49,** 904 (1936). See also Morse, Fisk and Schiff, *Phys. Rev.* **50,** 748 (1936).

‡ Fermi, *Phys. Rev.* **48,** 570 (1935).

§ Breit, Condon and Present, *Phys. Rev.* **50,** 825 (1936).

of the interaction laws for various kinds of particles so far as specifically nuclear forces are concerned.

This paper grew out of association at the 1936 summer symposium on theoretical physics of the University of Michigan. We wish to express here to Professor H. M. Randall our deep appreciation of the opportunity of working in the stimulating atmosphere of the Michigan laboratory.

10

Conservation of Isotopic Spin in Nuclear Reactions†‡

ROBERT K. ADAIR

The effects of the charge independence of nuclear forces on nuclear reaction experiments is discussed. It is pointed out that charge independence establishes relationships between cross sections for some reactions and results in forbidding certain other reactions.

It is well known§|| that if the forces between two nucleons are independent of their charge states and depend only on their space and spin coordinates, the wave function of a system of nucleons will be invariant under certain charge transformations. This symmetry is usually described in terms of a constant of motion, the isotopic spin. Important consequences of this invariance of nuclear structure with respect to rotation in isotopic spin have been discussed by Wigner¶†† and others.

There is no very conclusive evidence to establish the charge independence of nuclear forces. While the similarity of energy levels of mirror nuclei strongly suggests the equivalence of neutron–neutron forces and proton–proton forces,‡‡ there is not yet much

† *Phys. Rev.* **87**, 1041 (1952).
‡ Work supported by the AEC and the Wisconsin Alumni Research Foundation.
§ E. Wigner, *Phys. Rev.* **51**, 106 (1937).
|| N. Kemmer, *Proc. Camb. phil. Soc.* **34**, 354 (1938).
¶ E. Feenberg and E. Wigner, *Phys. Rev.* **51**, 95 (1937).
†† E. Wigner, *Phys. Rev.* **56**, 519 (1939).
‡‡ See, e.g. V. R. Johnson, *Phys. Rev.* **86**, 302 (1952).

information on the equality of the neutron–proton interaction and the forces between like nucleons. In particular, no definite conclusions have been derived from nucleon–nucleon scattering. Although the low energy scattering experiments are not in contradiction to a description in terms of charge independent forces,† it is not clear that this is true of higher energy measurements.‡

It appears to have been largely overlooked in recent work that information on the charge independence of nuclear forces may be obtained from observations on nuclear reactions. Charge independence has observable consequences in such reactions; in particular, it results in forbidding certain transitions which are allowed solely from considerations of spin and parity.

For the purpose of this discussion the third component T_3 of the isotopic spin T of a nucleus is defined, as usual, as the number of neutrons minus the number of protons in the nucleus divided by two. Systems having the same isotopic spin but different T_3 components form a set of multiplicity $2T+1$, which differ in energy only through Coulomb forces. On the assumption that it is a good quantum number, the value of T for any nucleus is easily determined by an examination of the binding energies of nuclei.§|| Generally the binding energy of light nuclei depends strongly upon T. Since there is close competition between low states with different values of T only in the $4n-2$ group of nuclei, most of the interesting applications of charge symmetry selection rules take place in reactions involving these systems. A typical example of such a system is the C^{14}, N^{14}, O^{14} triad. The isotopic spin function representing the ground state of N^{14} can be written as $\varphi'{}_0{}^0$, of C^{14}, $\varphi_1{}^1$, and of O^{14}, $\varphi_1{}^{-1}$, where the superscript represents the value of the T_3 component and the subscript the value of T. The $\varphi_1{}^0$ state will be an excited state of N^{14} displaced in energy from the O^{14} and C^{14} ground states by Coulomb forces and the neutron–proton mass difference. This appears to be the 2·3 MeV level of N^{14}.

† J. Schwinger, *Phys. Rev.* **78**, 135 (1950).
‡ R. S. Christian and H. P. Noyes, *Phys. Rev.* **79**, 85 (1950).
§ E. Feenberg and M. Phillips, *Phys. Rev.* **51**, 597 (1937).
|| Hornyak, Lauritsen, Morrison, and Fowler, *Revs. Modern Phys.* **22**, 291 (1950).

The transformation properties of isotopic spin are the same as for angular momentum. Since the assumption of charge independence is equivalent to requiring the conservation of total isotopic spin and conservation of charge insures the conservation of the T_3 component, isotopic spin selection rules are identical with those of angular momentum. An example of a reaction forbidden because of isotopic spin selection rules is the O^{16}(d,α) reaction to the 2·3 MeV state of N^{14}. Since the isotopic spin of O^{16}, the deuteron, and the alpha-particle are all zero, while the N^{14} level has an isotopic spin of one, the transition will not be allowed. Although this reaction has been investigated by several groups†‡§ under a variety of experimental conditions it has not been observed.

There are a number of other reactions which should be forbidden by analogous considerations. The C^{12}(d,α) reaction to the 1·7 MeV state of B^{10} should not be observed, as this level appears to be the $\varphi_1{}^0$ state of the Be^{10}, B^{10}, C^{10} triad of total isotopic angular momentum one. The C^{12}(d,α) reaction has not been investigated at energies which would excite the 1·7 MeV level. For similar reasons the Ne^{20}(d,α) reaction to states in F^{18} which have corresponding levels in O^{18} should be forbidden, as these transitions would also violate the conservation of isotopic spin. Middleton and Tai investigated|| the Ne^{20}(d,α) reaction and found a prominent group of alpha-particles which leave the F^{18} nucleus in a 1·05 MeV excited state. This is near the energy where one would expect the isotopic spin-one level corresponding to the ground state of O^{18}. According to the shell model F^{18} consists of a closed P-shell plus a neutron and a proton. There is close competition between $S_{\frac{1}{2}}$ and $D_{\frac{5}{2}}$ states in this region. Since the beta-decay of F^{18} to O^{18} is allowed, the ground state of F^{18} probably belongs to the spin one, isotopic spin zero, triplet $(S_{\frac{1}{2}})^2$ configuration. The spin zero, singlet $(S_{\frac{1}{2}})^2$ configuration with isotopic spin one, corresponding to the ground state of O^{18},

† A. Ashmore and J. F. Raffle, *Proc. Phys. Soc.* (*London*), **A64,** 754 (1950).
‡ Burrows, Powell, and Rotblat, *Proc. roy. Soc.* (*London*), **A209,** 478 (1951).
§ Craig, Donahue, and Jones (to be published); Van de Graaff, Sperduto, Buechner, and Enge, *Phys. Rev.* **86,** 966 (1952).
|| R. Middleton and C. T. Tai, *Proc. Phys. Soc.* (*London*), **A64,** 801 (1951).

would be somewhat higher and might be close to the lowest $(D_{\frac{5}{2}})^2$ level, probably with spin five and isotopic spin zero. It therefore seems possible that this reaction does not violate charge independence but that the 1·05 MeV state reached by this reaction is an isotopic spin-zero level very close in energy to the isotopic spin-one level corresponding to the ground state of O^{18}.

The inelastic scattering of deuterons and alpha-particles by Li^6, B^{10}, and N^{14} will be affected by charge independence selection rules. Since the ground states of these nuclei, and the deuteron or alpha-particle, all have an isotopic spin of zero, the low-lying isotopic spin-one levels of these nuclei should not be excited appreciably in the reactions.

It must be emphasized that, since the transition probability depends upon the square of the matrix element, reaction experiments are not a sensitive test of charge independence. A reaction cross section leading to a state consisting of a mixture of 10 per cent of a wave function to which the transition were allowed and 90 per cent of a wave function to which the transition were forbidden would result in a cross section only of the order of one per cent of a completely allowed reaction. Conversely, the selection rules should hold reasonably well even if nuclear forces are only approximately charge independent. It is therefore necessary to place rather small upper limits on the cross sections for forbidden reactions in order to restrict very severely the dependence of nuclear forces on the nuclear charge. Coulomb forces will couple states of different isotopic spin but their influence on the wave functions of light nuclei is probably small. However, one cannot then expect to detect differences in the specifically nuclear forces of the magnitude of the Coulomb forces.†

It is sometimes easier experimentally to place a small limit on level widths rather than on reaction cross sections. If elastic scattering is

† *Note added in proof*:—I wish to thank Dr. N. M. Kroll for pointing out to me that selection rules will obtain in self-conjugate nuclei from the less severe condition of charge symmetry of nuclear forces. [See also in this connection Lynne E. H. Trainor, *Phys. Rev.* **85**, 962 (1952).] The evidence for charge independence in reactions involving these systems then depends upon the conjoint existence of selection rules affecting isotopic spin-one states, and the existence and energy equivalence of isobaric members of the charge triplet.

the most probable process taking place, the scattering cross section at resonance is reasonably large and practically independent of the magnitude of the reaction matrix element. The width of the level is proportional to the square of the matrix element. There exist levels† in B^{10} between 4·5 and 6·5 MeV above the ground state which can break up through heavy particle emission only to the isotopic spin-zero combinations of Li^6 plus an alpha-particle, or Be^8 plus a deuteron. If the specifically nuclear forces are charge independent, such transitions from isotopic spin-one states will only occur through the difference in neutron and proton wave functions caused by Coulomb forces. This effect should be small, and isotopic spin-one states should have very small widths. It might be possible to observe such states and measure their widths by scattering alpha-particles from Li^6.

Isotopic spin-one states of Li^6 could be investigated in a similar manner by scattering deuterons from helium. The lowest isotopic spin-one level is available energetically but, as Fowler‡ has pointed out, probably has spin zero and even parity and cannot therefore decay into a deuteron and an alpha-particle. There should, however, exist a level of Li^6 of even parity, isotopic spin one, probably of total angular momentum two, about 6 MeV above the ground state, corresponding to the first excited state of He^6. This state is forbidden by charge conservation from breaking down to a deuteron and an alpha-particle and should, therefore, have a very small width. It is possible energetically for this state to break up into He^5 plus a proton. If this width is much larger than the scattering width the resonance will be damped out and not easily observed by bombarding alpha-particles with deuterons.

The effects of charge independence selection rules should be discernible in other interactions involving states in a compound nucleus. The resonant scattering of protons§ from Be^9 at 1·087 MeV bombarding energy has been interpreted|| in terms of the formation of a

† Fay Ajzenberg, *Phys. Rev.* **82,** 43 (1951).
‡ Quoted by R. B. Day and R. L. Walker, *Phys. Rev.* **85,** 582 (1952).
§ Thomas, Rubin, Fowler, and Lauritsen, *Phys. Rev.* **75,** 1612 (1949).
|| R. Cohen, Ph.D. Thesis, *Cal. Inst. Tech.* (1949).

compound state of spin zero and odd parity in the compound nucleus B^{10}. Although lower levels decay by both alpha-particle and deuteron emission this state does not appear to do so, though these transitions are not unfavoured by angular momentum considerations. This behaviour might be explained by presuming the isotopic spin of the state to be one, as in that case transitions to the isotopic spin-zero combinations of Li^6 and an alpha-particle or Be^8 and a deuteron would be forbidden.

If the forces between nucleons are independent of their charge, relationships will be established between transition probabilities to states of different charge belonging to the same isotopic spin multiplet. An example of this is the Be^9(d,p) and Be^9(d,n) reactions to the ground state of Be^{10} and the 1·7 MeV state of B^{10}, respectively. These states are the ${\varphi_1}^1$ and ${\varphi_1}^0$ components of a charge multiplet. Writing the initial charge wave function, representing the system consisting of Be^9 plus the deuteron, as ${I_{\frac{1}{2}}}^{\frac{1}{2}}$, and the nuclear isotopic spin functions as ${\tau_{\frac{1}{2}}}^{\frac{1}{2}}$ for the neutron and ${\tau_{\frac{1}{2}}}^{-\frac{1}{2}}$ for the proton, we can expand I in terms of φ and τ:

$$I^{\frac{1}{2}}_{\frac{1}{2}} = (2^{\frac{1}{2}}\varphi^1_1\tau^{-\frac{1}{2}}_{\frac{1}{2}} - \varphi^0_1\tau^{\frac{1}{2}}_{\frac{1}{2}})/\sqrt{3}$$

The ratio of the transition probabilities to Be^{10} and B^{10} will then be $({I_{\frac{1}{2}}}^{\frac{1}{2}}|{\varphi_1}^1{\tau_{\frac{1}{2}}}^{-\frac{1}{2}})^2/({I_{\frac{1}{2}}}^{\frac{1}{2}}|{\varphi_1}^0{\tau_{\frac{1}{2}}}^{\frac{1}{2}})^2$, which is equal to two, multiplied by the relative volumes of phase space available to the final systems. Uncertainties in the magnitude of Coulomb effects are sufficiently large so that this type of reaction probably cannot be used as a sensitive check on charge independence. However, the cross section relations may be useful in determining which excited states in systems like N^{14}, B^{10}, and F^{18}, correspond to the levels of isotopic spin one found in their isobaric neighbours.

I wish to thank Professor J. M. Luttinger, who suggested the possible importance of charge independence in nuclear reactions, for his interest in this discussion.

11

The Effect of Charge Symmetry on Nuclear Reactions†‡

NORMAN M. KROLL

AND

LESLIE L. FOLDY

For a charge symmetric nuclear Hamiltonian, the operator which changes neutrons into protons and protons into neutrons (charge parity operator) commutes with the Hamiltonian and is therefore a constant of the motion. Since the charge parity operator anticommutes with the "3" component of the total isotopic spin, for nuclei with $T_3 = 0$ (self-conjugate nuclei) the charge parity is a good quantum number and in the absence of degeneracy the eigenstates of such nuclei have either odd or even charge parity. This leads to strong selection rules in nuclear reactions involving self-conjugate nuclei in the initial and final states which may reasonably be invoked to explain recent experimental results on such reactions. Since states of even total isotopic spin have even charge parity and states of odd total isotopic spin have odd parity, the selection rules arising from charge symmetry often coincide with those of charge independence and in such cases a definitive test of the charge independence hypothesis by the use of these selection rules is impeded. Some other applications of the charge symmetry principle are discussed.

† *Phys. Rev.* **88**, 1177 (1952).
‡ This research was supported by the AEC and carried out while the authors were Visiting Scientists at Brookhaven National Laboratory during the summer of 1952.

THAT the neutron–neutron and proton–proton forces are equivalent apart from electromagnetic interactions is strongly indicated by the energy differences in the ground states and the general similarity of energy levels in the various mirror nuclei. It is the purpose of this note to point out that the assumption of such a charge symmetry of nuclear forces implies certain strong selection rules† which apply in a small but experimentally interesting group of nuclear reactions.

These selection rules arise in self-conjugate systems, that is, systems having equal numbers of neutrons and protons. For such systems under the assumption of charge symmetry the Hamiltonian is invariant under interchange of the neutron and proton space and spin coordinates. It is therefore possible to define an operator P which performs this interchange and can appropriately be called the charge parity operator.‡ It is clear that P is a constant of the motion with eigenvalues 1 and -1. The wave functions of the system can always be selected so as to be simultaneous eigenfunctions of the Hamiltonian and charge parity and therefore may be characterized as charge even or charge odd. Of course in the degenerate case where two states of opposite charge parity have the same energy a state of the system of this energy may be a superposition of charge even and charge odd states.

A formal exposition of the above described concept can be conveniently obtained in terms of the isotopic spin formalism. The operator P can be represented as a rotation in isotopic spin space of

† The selection rules referred to would be rigorous were it not for the neutron–proton mass difference and the fact that the electromagnetic interactions between nucleons are not charge symmetric. This limitation on the universality of the charge symmetry hypothesis sets practical limits on all derived consequences of the hypothesis. The fact that these nonsymmetric interactions are generally weak compared to specifically nuclear interactions for the cases of interest implies, however, that the selection rules obtained have considerable potency.

‡ The earliest reference to the existence of a good quantum number for self-conjugate nuclei associated with the charge parity operator appears to be in a paper of E. Feenberg and E. P. Wigner, *Phys. Rev.* **51**, 95 (1937). Recently the concept of charge parity was discussed by L. Trainor, *Phys. Rev.* **85**, 962 (1952), in connection with its application to the problem of electric dipole radiation from self-conjugate nuclei. It has the consequence here of forbidding electric dipole radiation in a transition between two states of the same charge parity.

180° about the 1 axis. That is, for a system of A particles the total isotopic spin $\mathbf{T}$ is given by

$$\mathbf{T} = \sum_{j=1}^{A} \boldsymbol{\tau}^{(j)}$$

A rotation of 180° about the 1 axis is then given by

$$P = \exp(i\pi T_1/2) = \prod_{j=1}^{A} \exp(i\pi\tau_1^{(j)}/2) = i^A \prod_{j=1}^{A} \tau_1^{(j)}$$

since $\exp(i\pi\tau_1^{(j)}/2) = i\pi\tau_1^{(j)}$.

The charge symmetry of the Hamiltonian implies that it is a symmetric function of the A particles whose dependence upon the isotopic spin† can be expressed in terms of $\boldsymbol{\tau}^{(i)}.\boldsymbol{\tau}^{(j)}$ and $\tau_3{}^{(i)}\tau_3{}^{(j)}$. Since $[\boldsymbol{\tau}^{(i)}.\boldsymbol{\tau}^{(j)}, P] = [\tau_3^{(i)}\tau_3^{(j)}, P] = 0$, it is clear that P commutes with the Hamiltonian and is a constant of the motion. This property is not, in fact, restricted to self-conjugate systems. On the other hand, nontrivial applications of the concept are in fact so restricted. This arises from the fact that $[P, T_3] = 2PT_3$ that is, P and T_3 anti-commute. Thus a system can be in a simultaneous eigenstate of P and T_3 only for states of $T_3 = 0$, which corresponds to the case of equal numbers of neutrons and protons. The physical meaning of the above is quite obvious. P simply transforms protons into neutrons and neutrons into protons. Since T_3 is simply $\frac{1}{2}(N-Z)$, the application of P to an eigenstate of T_3 with eigenvalue t_3 simply changes it to an eigenstate of T_3 with an eigenvalue $-t_3$.

We note that $P^2 = (-1)^A$ so that for even A, P has eigenvalues $+1$ and -1. A further property of P which is of interest for the discussion to follow is the fact that $[T^2, P] = 0$, so that T^2 and P have simultaneous eigenstates. Furthermore, a $T_3 = 0$ eigenstate of T^2 with eigenvalue $t(t+1)$ must, in fact, be an eigenstate of charge parity and odd or even as t is odd or even. This latter result follows from the fact that under a rotation in isotopic spin space, the eigenfunctions of T^2 corresponding to a given t must transform like

† This form of the Hamiltonian includes (in addition to the equality of nn and pp forces) the assertion that the interaction energy between neutrons and protons involves the neutron and proton space and spin coordinates symmetrically. It is clear that usually asserted consequences of charge symmetry with respect to the mirror nuclei are valid only when this additional assumption is included.

the spherical harmonic of order t under the homologous rotation in coordinate space.

The principal application of the concept of charge parity arises in the case of reactions between nuclei for which the incident and product nuclei are individually self-conjugate. In just these cases, the initial and final states of the system will be eigenstates of charge parity. From the fact that P is a constant of the motion, the initial and final states must both be charge even or charge odd. As a specific example one might consider the $O^{16}(d,\alpha)N^{14}$ reactions recently investigated by various research groups.† In this case one finds prominent α-particle groups corresponding to the ground state and various excited states of N^{14} but none corresponding to the 2·3 MeV excited state. A reasonable interpretation of this result is simply that this particular state has charge parity opposite to that of the states observed.‡ It is almost certain that the ground state of N^{14} is charge even, while if the nuclear forces are only approximately charge independent one would expect a low-lying charge odd state.

In order to make clear the relevancy of having both the incident and product nuclei in charge conjugate states one might note that a reaction like $O^{16}(d,p)O^{17*}$ will never be forbidden by charge parity conservation in spite of the fact that the initial state is charge even. The requirement that the final state be charge even as well simply implies that the reactions $O^{16}(d,p)O^{17*}$ and $O^{16}(d,n)F^{17*}$ occur with equal probability, where the final states of O^{17*} and F^{17*} are mirror states.

On the other hand, the observation of nuclear resonances associated with compound nucleus formation can be affected by charge parity considerations when only the product nuclei or only the

† Ashmore and Raffle, *Proc. Phys. Soc. (London)*, **A64**, 754 (1950); Burrows, Powell, and Rotblat, *Proc. roy. Soc. (London)*, **A209**, 478 (1951); Van de Graaff, Sperderto, Beuchner, and Enge, *Phys. Rev.* **86**, 966 (1952).

‡ *Note added in proof*: E. Feenberg (private communication) has pointed out that the negative results obtained by the Van de Graaff, etc. group using 2·1 MeV deuterons might possibly be attributed entirely to the effect of the Coulomb barrier on the outgoing α-particles combined with the effects of angular momentum and space parity conservation associated with the probable assignment of spin zero and positive space parity to the N^{14} state. These effects appear, however, to be unimportant in the other experiments, which made use of higher energy deuterons.

incident nuclei are self conjugate. For example, the observation of a resonance in the reaction $C^{12}(d,p)C^{13}$ due to the formation of an excited state of N^{14} implies that the state of N^{14} is charge even, although in this case the product nuclei are not self conjugate.

The existence of charge parity has certain interesting connections with the problem of charge independence of nuclear forces. As has been pointed out by Adair,† the assumption of charge independence also implies selection rules in nuclear reactions, which, while including those implied by charge symmetry, are considerably more far reaching. This is a consequence of the fact that total isotopic spin as well as charge parity must be conserved. It is unfortunate, however, that the role of isotopic spin is obscured in many reactions by the coincidence of its predictions with those of charge parity. It has, for example, been proposed that the 2·3 MeV N^{14} state is a member of an isotopic spin multiplet with $T = 1$, while the ground state is $T = 0$. It is important to realize that it is then not possible for these states to have the same charge parity. It follows that it is difficult to find suitable reactions involving these states ($T = 0$ and $T = 1$) which are allowed by charge parity conservation and prohibited by isotopic spin conservation. Thus the observed prohibition of the previously mentioned reaction, $O^{16}(d,\alpha)N^{14*}$, yields no direct information on the strength of np forces as compared with nn and pp forces.‡

Another application of the charge parity operator is in determining the terms which may be admixed in a particular state of a nucleus. Thus consider the case of a neutron and a proton both in the same p orbital in a nucleus such as in the case of Li^6. The terms which may be constructed from this configuration with total angular momentum unity are $^3S_1$, $^3D_1$, $^1P_1$, and $^3P_1$. These terms will not be all

† R. K. Adair, *Phys. Rev.* **87,** 1044 (1952) (see p. 202 of this book). The authors wish to acknowledge the fact that their thoughts on the substance of this article were initiated by Adair's work, and to thank V. F. Weisskopf for calling it to their attention.

‡ On the other hand, if one assumes the nuclear forces to be approximately charge independent than this result does contribute to the identification of the symmetry character of the state. The energy of the state is then entirely consistent with charge independence.

admixed (except again in the case of a degeneracy), since the first three of these have even charge parity and the last has odd charge parity.

Our final application of charge parity will be to the problem of the β-decay† of O^{14} to the 2·3 MeV excited state of N^{14}. Definite identification of the excited state as belonging to the same isotopic spin triplet to which the ground state of O^{14} belongs (under the assumption of charge independence) would confirm the angular momentum assignment of this state as $I = 0$ and thus allow an estimation of the β-decay coupling constant associated with Fermi selection rules. Blatt‡ recently pointed out that Adair's interpretation of the aforementioned $O^{16}(d,\alpha)N^{14*}$ result would indeed confirm this identification. Furthermore, the fact that the transition is super-allowed implies that the matrix element has its maximum value (two).§ The relevance of charge parity arises from the fact that it weakens the dependence of these conclusions on the assumption of charge independence and, therefore, strengthens the conclusions drawn. Thus if one considers the deviation from charge independence as a perturbation, one notes that there are no nearby states of N^{14} with which the $T = 1$ state can mix. To elaborate, the energy level diagram of O^{14} indicates a separation of about 6 MeV between the ground state and first excited state. The several states in the vicinity of the $T = 1$ state of N^{14} must, therefore, all be $T = 0$ states. If one now considers the effect of a deviation from charge independence on the $T = 1$ state of N^{14} it is clear that the charge even character of $T = 0$ states prohibits the mixing (which might otherwise be large) of these states with the $T = 1$ state. Thus the expectation that the matrix element of the $O^{14} \rightarrow N^{14*}$ transition is two is only weakly affected.

It is a pleasure to thank Professor V. F. Weisskopf for stimulating and contributory discussion.

† Sherr, Muether, and White, *Phys. Rev.* **75**, 282 (1949).
‡ J. Blatt (private communication).
§ G. L. Trigg, *Phys. Rev.* **86**, 506 (1952).

12

On the Interaction of Elementary Particles†

HIDEKI YUKAWA

1. Introduction

AT the present stage of the quantum theory little is known about the nature of interaction of elementary particles. Heisenberg considered the interaction of "Platzwechsel" between the neutron and the proton to be of importance to the nuclear structure.‡

Recently Fermi treated the problem of β-disintegration on the hypothesis of "neutrino".§ According to this theory, the neutron and the proton can interact by emitting and absorbing a pair of neutrino and electron. Unfortunately the interaction energy calculated on such assumption is much too small to account for the binding energies of neutrons and protons in the nucleus.||

To remove this defect, it seems natural to modify the theory of Heisenberg and Fermi in the following way. The transition of a heavy particle from neutron state to proton state is not always accompanied by the emission of light particles, i.e. a neutrino and an electron, but the energy liberated by the transition is taken up sometimes by another heavy particle, which in turn will be transformed from proton state into neutron state. If the probability of occurrence

† *Proc. Math. Soc. Japan*, **17,** 48 (1935).

‡ W. Heisenberg, *Z. Phys.* **77,** 1 (1932); **78,** 156 (1932); **80,** 587(1933) (see pp. 144, 155 of this book). We shall denote the first of them by I.

§ E. Fermi, *ibid.* **88,** 161 (1934).

|| Ig. Tamm, *Nature*, **133,** 981 (1934); D. Iwanenko, *ibid.* **981** (1934).

of the latter process is much larger than that of the former, the interaction between the neutron and the proton will be much larger than in the case of Fermi, whereas the probability of emission of light particles is not affected essentially.

Now such interaction between the elementary particles can be described by means of a field of force, just as the interaction between the charged particles is described by the electromagnetic field. The above considerations show that the interaction of heavy particles with this field is much larger than that of light particles with it.

In the quantum theory this field should be accompanied by a new sort of quantum, just as the electromagnetic field is accompanied by the photon.

In this paper the possible natures of this field and the quantum accompanying it will be discussed briefly and also their bearing on the nuclear structure will be considered.

Besides such an exchange force and the ordinary electric and magnetic forces there may be other forces between the elementary particles, but we disregard the latter for the moment.

Fuller account will be made in the next paper.

2. Field Describing the Interaction

In analogy with the scalar potential of the electromagnetic field, a function $U(x, y, z, t)$ is introduced to describe the field between the neutron and the proton. This function will satisfy an equation similar to the wave equation for the electromagnetic potential.

Now the equation

$$\left(\nabla^2 - \frac{1}{c^2}\frac{\partial^2}{\partial t^2}\right) U = 0 \tag{1}$$

has only static solution with central symmetry $1/r$, except the additive and the multiplicative constants. The potential of force between the neutron and the proton should, however, not be of Coulomb type, but decrease more rapidly with distance. It can be expressed, for example, by

$$+ \text{ or } -g^2 \exp(-\lambda r)/r \tag{2}$$

where g is a constant with the dimension of electric charge, i.e. $cm^{\frac{3}{2}}\ sec^{-1}\ g^{\frac{1}{2}}$ and λ with the dimension cm^{-1}.

Since this function is a static solution with central symmetry of the wave equation

$$\left(\nabla^2 - \frac{1}{c^2}\frac{\partial^2}{\partial t^2} - \lambda^2\right)U = 0 \tag{3}$$

let this equation be assumed to be the correct equation for U in vacuum. In the presence of the heavy particles, the U-field interacts with them and causes the transition from neutron state to proton state.

Now, if we introduce the matrices†

$$\tau_1 = \begin{pmatrix} 0 & 1 \\ 1 & 0 \end{pmatrix}, \quad \tau_2 = \begin{pmatrix} 0 & -i \\ i & 0 \end{pmatrix}, \quad \tau_3 = \begin{pmatrix} 1 & 0 \\ 0 & -1 \end{pmatrix}$$

and denote the neutron state and the proton state by $\tau_3 = 1$ and $\tau_3 = -1$ respectively, the wave equation is given by

$$\left\{\nabla^2 - \frac{1}{c^2}\frac{\partial^2}{\partial t^2} - \lambda^2\right\}U = -4\pi g\tilde{\Psi}\frac{\tau_1 - i\tau_2}{2}\Psi \tag{4}$$

where Ψ denotes the wave function of the heavy particles, being a function of time, position, spin as well as τ_3, which takes the value either 1 or -1.

Next, the conjugate complex function $\tilde{U}(x, y, z, t)$, satisfying the equation

$$\left\{\nabla^2 - \frac{1}{c^2}\frac{\partial^2}{\partial t^2} - \lambda^2\right\}\tilde{U} = -4\pi g\tilde{\Psi}\frac{\tau_1 + i\tau_2}{2}\Psi \tag{5}$$

is introduced, corresponding to the inverse transition from proton to neutron state.

Similar equation will hold for the vector function, which is the analogue of the vector potential of the electromagnetic field. However, we disregard it for the moment, as there is no correct relativistic theory for the heavy particles. Hence simple non-relativistic wave equation neglecting spin will be used for the heavy particle, in the following way

† Heisenberg, *loc. cit.* I.

$$\left\{\frac{\hbar^2}{4}\left(\frac{1+\tau_s}{M_n}+\frac{1-\tau_3}{M_p}\right)\nabla^2+i\hbar\frac{\partial}{\partial t}-\frac{1+\tau_3}{2}M_nc^2-\frac{1-\tau_3}{2}M_pc^2-\right.$$
$$\left.-g\left(\tilde{U}\frac{\tau_1-i\tau_2}{2}+U\frac{\tau_1+i\tau_2}{2}\right)\right\}\Psi=0 \tag{6}$$

where $\hbar$ is Planck's constant divided by 2π and M_n, M_p are the masses of the neutron and the proton respectively. The reason for taking the negative sign in front of g will be mentioned later.

The equation (6) corresponds to the Hamiltonian

$$H=\left(\frac{1+\tau_3}{4M_n}+\frac{1-\tau_3}{4M_p}\right)\boldsymbol{p}^2+\frac{1+\tau_3}{2}M_nc^2+\frac{1-\tau_3}{2}M_pc^2+$$
$$+g\left(\tilde{U}\frac{\tau_1-i\tau_2}{2}+U\frac{\tau_1+i\tau_2}{2}\right) \tag{7}$$

where $\boldsymbol{p}$ is the momentum of the particle. If we put $M_nc^2-M_pc^2=D$ and $M_n+M_p=2M$, the equation (7) becomes approximately

$$H=\frac{\boldsymbol{p}^2}{2M}+\frac{g}{2}\{\tilde{U}(\tau_1-i\tau_2)+U(\tau_1+i\tau_2)\}+\frac{D}{2}\tau_3 \tag{8}$$

where the constant term Mc^2 is omitted.

Now consider two heavy particles at points (x_1, y_1, z_1) and (x_2, y_2, z_2) respectively and assume their relative velocity to be small. The fields at (x_1, y_1, z_1) due to the particle at (x_2, y_2, z_2) are, from (4) and (5),

$$\left.\begin{aligned} U(x_1,y_1,z_1)&=\frac{g}{r_{12}}\exp(-\lambda r_{12})\tfrac{1}{2}(\tau_1^2-i\tau_2^2)\\ \text{and}\qquad \tilde{U}(x_1,y_1,z_1)&=\frac{g}{r_{12}}\exp(-\lambda r_{12})\tfrac{1}{2}(\tau_1^2+i\tau_2^2)\end{aligned}\right\} \tag{9}$$

where $(\tau_1^1, \tau_2^1, \tau_3^1)$ and $(\tau_1^2, \tau_2^2, \tau_3^2)$ are the matrices relating to the first and the second particles respectively, and r_{12} is the distance between them.

Hence the Hamiltonian for the system is given, in the absence of the external fields, by

$$H = \frac{\boldsymbol{p}_1^2}{2M} + \frac{\boldsymbol{p}_2^2}{2M} + \frac{g^2}{4}\{(\tau_1' - i\tau_2')(\tau_1^2 + i\tau_2^2) + (\tau_1^1 + i\tau_2^1)(\tau_1^2 - i\tau_2^2)\} \times$$
$$\times \exp(-\lambda r_{12})/r_{12} + (\tau_3^1 + \tau_3^2)D$$
$$= \frac{\boldsymbol{p}_1^2}{2M} + \frac{\boldsymbol{p}_2^2}{2M} + \frac{g^2}{2}(\tau_1^1\tau_1^2 + \tau_2^1\tau_2^2)\exp(-\lambda r_{12})/r_{12} +$$
$$+ (\tau_3^1 + \tau_3^2)D \qquad (10)$$

where $\boldsymbol{p}_1$, $\boldsymbol{p}_2$ are the momenta of the particles.

This Hamiltonian is equivalent to Heisenberg's Hamiltonian (1),† if we take for "Platzwechselintegral"

$$J(r) = -g^2 \exp(-\lambda r)/r \qquad (11)$$

except that the interaction between the neutrons and the electrostic repulsion between the protons are not taken into account. Heisenberg took the positive sign for $J(r)$, so that the spin of the lowest energy state of H^2 was O, whereas in our case, owing to the negative sign in front of g^2, the lowest energy state has the spin 1, which is required from the experiment (cf. Part 1, pp. 86, 90).

Two constants g and λ appearing in the above equations should be determined by comparison with experiment. For example, using the Hamiltonian (10) for heavy particles, we can calculate the mass defect of H^2 and the probability of scattering of a neutron by a proton provided that the relative velocity is small compared with the light velocity.‡

Rough estimation shows that the calculated values agree with the experimental results, if we take for λ the value between 10^{12} cm^{-1} and 10^{13} cm^{-1} and for g a few times of the elementary charge $\boldsymbol{e}$, although no direct relation between g and $\boldsymbol{e}$ was suggested in the above considerations.

3. Nature of the Quanta Accompanying the Field

The U-field above considered should be quantized according to the general method of the quantum theory. Since the neutron and the

† Heisenberg, I.

‡ These calculations were made previously, according to the theory of Heisenberg, by Mr. Tomonaga, to whom the writer owes much. A little modification is necessary in our case. Detailed accounts will be made in the next paper.

proton both obey Fermi's statistics, the quanta accompanying the U-field should obey Bose's statistics and the quantization can be carried out on the line similar to that of the electromagnetic field.

The law of conservation of the electric charge demands that the quantum should have the charge either $+e$ or $-e$. The field quantity U corresponds to the operator which increases the number of negatively charged quanta and decreases the number of positively charged quanta by one respectively. $\tilde{U}$, which is the complex conjugate of U, corresponds to the inverse operator.

Next, denoting

$$p_x = -i\hbar\frac{\partial}{\partial x}, \text{ etc.,} \qquad W = i\hbar\frac{\partial}{\partial t}, \qquad m_U c = \lambda\hbar$$

the wave equation for U in free space can be written in the form

$$\left\{p_x^2 + p_y^2 + p_z^2 - \frac{W^2}{c^2} + m_U c^2\right\} U = 0 \tag{12}$$

so that the quantum accompanying the field has the proper mass $m_U = \lambda\hbar/c$. Assuming $\lambda = 5\times 10^{12}\ \text{cm}^{-1}$, we obtain for m_U a value 2×10^2 times as large as the electron mass. As such a quantum with large mass and positive or negative charge has never been found by the experiment, the above theory seems to be on a wrong line. We can show, however, that, in the ordinary nuclear transformation, such a quantum cannot be emitted into outer space.

Let us consider, for example, the transition from a neutron state of energy W_n to a proton state of energy W_p, both of which include the proper energies. These states can be expressed by the wave functions

$$\Psi_n(x, y, z, t, 1) = u(x, y, z)\exp(-iW_n t/\hbar), \qquad \Psi_n(x, y, z, t, -1) = 0$$

and

$$\Psi_p(x, y, z, t, 1) = 0$$

$$\Psi_p(x, y, z, t, -1) = v(x, y, z)\exp(-iW_p t/\hbar)$$

so that, on the right hand side of the equation (4), the term

$$-4\pi g\tilde{v}u\exp[-it(W_n - W_p)/\hbar]$$

appears.

Putting $U = U'(x, y, z) \exp(i\omega t)$, we have from (4)

$$\left\{\nabla^2 - \left(\lambda^2 - \frac{\omega^2}{c^2}\right)\right\} U' = -4\pi g \tilde{v} u \tag{13}$$

where $\omega = W_n - W_l/\hbar$. Integrating this, we obtain a solution

$$U'(r) = g \iiint \frac{\exp(-\mu|r-r'|)}{|r-r'|} \tilde{v}(r') u(r') \, dv' \tag{14}$$

where $\mu = (\lambda^2 - \omega^2/c^2)^{\frac{1}{2}}$.

If $\lambda > |\omega|/c$ or $m_U c^2 > |W_n - W_p|$, μ is real and the function $J(r)$ of Heisenberg has the form $-g^2 \exp(-\mu r)/r$, in which μ, however, depends on $|W_n - W_p|$, becoming smaller and smaller as the latter approaches $m_U c^2$. This means that the range of interaction between a neutron and a proton increases as $|W_n - W_p|$ increases.

Now the scattering (elastic or inelastic) of a neutron by a nucleus can be considered as the result of the following double process: the neutron falls into a proton level in the nucleus and a proton in the latter jumps to a neutron state of positive kinetic energy, the total energy being conserved throughout the process. The above argument, then, shows that the probability of scattering may in some case increase with the velocity of the neutron.

According to the experiment of Bonner,† the collision cross section of the neutron increases, in fact, with the velocity in the case of lead whereas it decreases in the case of carbon and hydrogen, the rate of decrease being slower in the former than in the latter. The origin of this effect is not clear, but the above considerations do not, at least, contradict it. For, if the binding energy of the proton in the nucleus becomes comparable with $m_U c^2$, the range of interaction of the neutron with the former will increase considerably with the velocity of the neutron, so that the cross section will decrease slower in such case than in the case of hydrogen, i.e. free proton. Now the binding energy of the proton in C^{12}, which is estimated from the difference of masses of C^{12} and B^{11}, is

$$12{\cdot}0036 - 11{\cdot}0110 = 0{\cdot}9926$$

This corresponds to a binding energy 0·0152 in mass unit, being

† T. W. Bonner, *Phys. Rev.* **45**, 606 (1934).

thirty times the electron mass. Thus in the case of carbon we can expect the effect observed by Bonner. The arguments are only tentative, other explanations being, of course, not excluded.

Next if $\lambda < |\omega|/c$ or $m_U c^2 < |W_n - W_p|$, μ becomes pure imaginary and U expresses a spherical undamped wave, implying that a quantum with energy greater than $m_U c^2$ can be emitted in outer space by the transition of the heavy particle from neutron state to proton state, provided that $|W_n - W_p) > m_U c^2$.

The velocity of U-wave is greater but the group velocity is smaller than the light velocity c, as in the case of the electron wave.

The reason why such massive quanta, if they ever exist, are not yet discovered may be ascribed to the fact that the mass m_U is so large that condition $|W_n - W_p| > m_U c^2$ is not fulfilled in ordinary nuclear transformation.

4. Theory of β-Disintegration

Hitherto we have considered only the interaction of U-quanta with heavy particles. Now, according to our theory, the quantum emitted when a heavy particle jumps from a neutron state to a proton state, can be absorbed by a light particle which will then in consequence of energy absorption rise from a neutrino state of negative energy to an electron state of positive energy. Thus an anti-neutrino and an electron are emitted simultaneously from the nucleus. Such intervention of a massive quantum does not alter essentially the probability of β-disintegration, which has been calculated on the hypothesis of direct coupling of a heavy particle and a light particle, just as, in the theory of internal conversion of γ-ray, the intervation of the proton does not affect the final result.† Our theory, therefore, does not differ essentially from Fermi's theory.

Fermi considered that an electron and a neutrino are emitted simultaneously from the radioactive nucleus, but this is formally equivalent to the assumption that a light particle jumps from a neutrino state of negative energy to an electron state of positive energy.

† H. A. Taylor and N. F. Mott, *Proc. Roy. Soc.* **A138**, 665 (1932).

For, if the eigenfunctions of the electron and the neutrino be ψ_k, φ_k respectively, where $k = 1, 2, 3, 4$, a term of the form

$$-4\pi g' \sum_{k=1}^{4} \tilde{\psi}_k \varphi_k \tag{15}$$

should be added to the right hand side of the equation (5) for $\tilde{U}$, where g' is a new constant with the same dimension as g.

Now the eigenfunctions of the neutrino state with energy and momentum just opposite to those of the state φ_k is given by $\varphi'_k = -\delta_{kl}\tilde{\varphi}_l$ and conversely $\varphi_k = \delta_{kl}\tilde{\varphi}'_l$, where

$$\delta = \begin{pmatrix} 0 & -1 & 0 & 0 \\ 1 & 0 & 0 & 0 \\ 0 & 0 & 0 & 1 \\ 0 & 0 & -1 & 0 \end{pmatrix},$$

so that (15) becomes

$$-4\pi g' \sum_{k,l=1}^{4} \tilde{\psi}_k \delta_{kl} \tilde{\varphi}'_l \tag{16}$$

From equations (13) and (15), we obtain for the matrix element of the interaction energy of the heavy particle and the light particle an expression

$$gg' \int \cdots \int \tilde{v}(\boldsymbol{r}_1) u(\boldsymbol{r}_1) \sum_{k=1}^{4} \psi_k(\boldsymbol{r}_2) \varphi_k(\boldsymbol{r}_2) \frac{\exp(-\lambda r_{12})}{r_{12}} \, dv_1 \, dv_2 \tag{17}$$

corresponding to the following double process: a heavy particle falls from the neutron state with the eigenfunction $u(\boldsymbol{r})$ into the proton state with the eigenfunction $v(\boldsymbol{r})$ and simultaneously a light particle jumps from the neutrino state $\varphi_k(\boldsymbol{r})$ of negative energy to the electron state $\psi_k(\boldsymbol{r})$ of positive energy. In (17) λ is taken instead of μ, since the difference of energies of the neutron state and the proton state, which is equal to the sum of the upper limit of the energy spectrum of β-rays and the proper energies of the electron and the neutrino, is always small compared with $m_U c^2$.

As λ is much larger than the wave numbers of the electron state and the neutrino state, the function $\exp(-\lambda r_{12})/r_{12}$ can be regarded approximately as a δ-function multiplied by $4\pi/\lambda^2$ for the integrations with respect to x_2, y_2, z_2. The factor $4\pi/\lambda^2$ comes from

$$\iiint \frac{\exp(-\lambda r_{12})}{r_{12}} \, dv_2 = \frac{4\pi}{\lambda^2}$$

Hence (17) becomes

$$\frac{4\pi g g'}{\lambda^2} \iiint \tilde{v}(\mathbf{r}) u(\mathbf{r}) \sum_k \tilde{\psi}_k(\mathbf{r}) \varphi_k(\mathbf{r}) \, dv \tag{18}$$

or by (16)

$$\frac{4\pi g g'}{\lambda^2} \iiint \tilde{v}(\mathbf{r}) u(\mathbf{r}) \sum_{k,l} \tilde{\psi}(\mathbf{r}) \delta'_{kl} \tilde{\varphi}'_l(\mathbf{r}) \, dv \tag{19}$$

which is the same as the expression (21) of Fermi, corresponding to the emission of a neutrino and an electron of positive energy states $\varphi'_k(\mathbf{r})$ and $\psi_k(\mathbf{r})$, except that the factor $4\pi g g'/\lambda^2$ is substituted for Fermi's g.

Thus the result is the same as that of Fermi's theory, in this approximation, if we take

$$\frac{4\pi g g'}{\lambda^2} = 4 \times 10^{-50} \text{ cm}^3 \text{ erg}$$

from which the constant g' can be determined. Taking, for example, $\lambda = 5 \times 10^{12}$ and $g = 2 \times 10^{-9}$, we obtain $g' \simeq 4 \times 10^{-17}$, which is about 10^{-8} times as small as g.

This means that the interaction between the neutrino and the electron is much smaller than that between the neutron and the proton so that the neutrino will be far more penetrating than the neutron and consequently more difficult to observe. The difference of g and g' may be due to the difference of masses of heavy and light particles.

5. Summary

The interaction of elementary particles are described by considering a hypothetical quantum which has the elementary charge and the proper mass and which obeys Bose's statistics. The interaction of such a quantum with the heavy particle should be far greater than that with the light particle in order to account for the large interaction of the neutron and the proton as well as the small probability of β-disintegration.

Such quanta, if they ever exist and approach the matter close enough to be absorbed, will deliver their charge and energy to the latter. If, then, the quanta with negative charge come out in excess, the matter will be charged to a negative potential.

These arguments, of course, of merely speculative character, agree with the view that the high speed positive particles in the cosmic rays are generated by the electrostatic field of the earth, which is charged to a negative potential.†

The massive quanta may also have some bearing on the shower produced by cosmic rays.

In conclusion the writer wishes to express his cordial thanks to Dr. Y. Nishina and Prof. S. Kikuchi for the encouragement throughout the course of the work.

† G. H. Huxley, *Nature*, **134,** 418, 571 (1934); Johnson, *Phys. Rev.* **45,** 569 (1934).

13

Range of Nuclear Forces in Yukawa's Theory†

G. C. WICK

FOUR years ago, Yukawa, in an attempt to develop a relativistic theory of the interaction of heavy particles in nuclei, was led to predict the existence of charged particles of mass intermediate between those of the electron and the proton.‡

In view of the great interest and hope raised by the striking discovery in cosmic rays of particles having just the desired mass, which one is naturally tempted to identify with Yukawa's particles, it may be desirable to have a derivation as elementary as possible of the fundamental relation:

$$\rho = \frac{h}{mc} \tag{1}$$

where ρ is range of the nuclear forces, h is Planck's constant, m is the mass of the "heavy electron", c is the velocity of light, which led Yukawa to his remarkable prediction.

It may perhaps be of interest, therefore, to point out that the meaning of relation (1) may be simply illustrated by an argument based on Heisenberg's Uncertainty Principle, in close analogy to Bohr's discussion of Gamow's formula and other related problems.

The argument runs as follows: in Yukawa's theory the interaction between heavy particles is carried by the semi-heavy particles, by means of simple emission and absorption processes (much in the

† *Nature*, **142**, 994 (1938).

‡ H. Yukawa, *Proc. Phys.-Math. Soc. Japan*, **17**, 48 (1935) (see p. 214 of this book), see also H. Fröhlich, W. Heitler and N. Kemmer, *Proc. Roy. Soc.*, **A166**, 154 (1938), and several other papers quoted there.

same way as the relativistic interaction between two electrons can be described in terms of emission and absorption of light quanta); these are not, of course, actual emission and absorption processes, which would be contrary to the energy principle; they are called, therefore, virtual transitions. Let us see, however, a little closer how it comes about that the energy principle is respected. One might try to show that this is not so by setting up some device which could "see" the heavy electron whilst it is travelling from one heavy particle to the other. In this case the energy principle can only be saved, as usual, if the uncontrollable energy exchange involved in the operation of the device is so large as to cover the energy excess actually observed, which is at least mc^2. Now the time t employed by the Yukawa particle in travelling from one heavy particle to the other is at least r/c, where r is the distance between the heavy particles. The time of operation of the device must on the other hand be smaller than t (otherwise the system will react as a whole, and the device will not be able to detect the presence of the individual Yukawa particle), but it need not be essentially smaller than this. We see, therefore, that the energy uncertainty will be, at most:

$$\Delta E \sim hc/r$$

The condition:

$$\Delta E > mc^2$$

actually gives the distance (1) as the limit up to which virtual transitions can make themselves felt without contradiction of the energy principle. It may be remarked that by assuming a velocity of the intermediate particle smaller than c, it is only possible to reduce the energy uncertainty further, so that the consideration of relativistic velocities actually gives the optimum conditions or the upper limit to which the interaction may extend.

I am very glad to express my thanks to Professor N. Bohr for his kind interest and the Fondazione Volta of the C.d.R. for a grant enabling me to stay in Copenhagen.

14

The Decay of Negative Mesotrons in Matter†

E. FERMI, E. TELLER

AND

V. WEISSKOPF

IN a recent experiment Conversi, Pancini, and Piccioni‡ observed separately the behaviour of positive and negative mesotrons coming to rest in iron or in graphite. They find that in iron the disintegration electrons are observed only for positive mesotrons. This was indeed to be expected§ because negative mesotrons after being slowed down can approach the nuclei and disappear by nuclear interactions. If, on the other hand, graphite is used for stopping the mesotrons, delayed disintegration electrons are observed to be about equally abundant for positive and negative mesotrons. This is in sharp disagreement with current expectations and seems to indicate that the interaction of mesotrons with nucleons according to the conventional schemes is many orders of magnitude weaker than usually assumed. The disappearance of a negative mesotron can be analysed into a process of approach of the mesotron to the nucleus and the

† *Phys. Rev.* **71**, 314 (1947).

‡ M. Conversi, E. Pancini and O. Piccioni, *Phys. Rev.* **71**, 209 (1947).

§ S. Tomonaga and G. Araki, *Phys. Rev.* **58**, 90 (1940).

process of capture by short range interaction of the mesotron and the nucleons.

The slowing down of mesotrons to an energy of about 2000 eV takes place according to the conventional theory. In estimating the energy loss for lower energies we have considered energy exchange with electrons and radiation. We consider the electrons as a degenerate gas with a maximum velocity v_0 and assume that the velocity V of the mesotron is small compared to v_0. Then the energy loss to the electrons per unit time is of the order of magnitude $e^4m^2T/(\hbar^3\mu)$. Here m and μ are the masses of the electrons and the mesotrons, respectively, and T is the kinetic energy of the mesotron. This formula allows losses of energy even when the total energy is negative (mesotron bound to an atom), and is valid as long as the mesotron moves outside the K orbit. At closer distances the formula will be somewhat modified and at the lowest energies loss by radiation will predominate. The mesotron reaches its lowest orbit around the nucleus in most solids in not more than 10^{-12} sec. This orbit is 200 times smaller than the radius for the K shell, which is for carbon about 10 times the nuclear radius and for iron about twice the nuclear radius. After reaching this orbit the mesotron can be found within the nucleus with a probability of 1/1000 in the case of carbon and a probability 1/10 in the case of iron.

According to the conventional mesotron theories, one will have to assume that the capture now proceeds according to one of the following schemes:

$$p+\mu^- = n+h\nu$$

$$X+\mu^- = N+Y$$

Here P and N stand for proton and neutron, μ signifies the mass of the mesotron, $h\nu$ is a light quantum, and X and Y stand for initial and residual nuclei in the capture process. The first calculation of these processes for a special form of mesotron interaction is due to Kobayasi and Okayama, and Sakata and Tanikawa.† The results depend to some extent upon the spin of the mesotron and the form

† Kobayasi and Okayama, *Proc. Phys.-Math. Soc. Japan*, **21**, 1 (1939); Sakata and Tanikawa, *Proc. Phys.-Math. Soc. Japan*, **21**, 58 (1939).

of the interaction assumed. For example, in the case of pseudoscalar mesotrons with an interaction energy given by

$$\sum_i (\hbar/\mu c)\,(4\pi)^{\frac{1}{2}} g\tau_i \int (\psi^* \sigma \psi)\,\mathrm{grad}\,\phi_i$$

(ψ is the wave function of the nucleons, ϕ_i of the mesotrons, μ is the mesotron mass, the index i refers to the charge, and τ is the isotopic spin operator), one obtains for the time of capture by process (1) for a mesotron already captured in its lowest orbit 10^{-18} and 10^{-20} sec in carbon and iron, respectively. Process (2) is likely to lead to 10 times shorter lives. This is negligible compared to the life of a negative mesotron of 2×10^{-6} sec.

The experimental result† leads to the conclusion that the time of capture from the lowest orbit of carbon is not less than the time of natural decay, that is, about 10^{-6} sec. This is in disagreement with the previous estimate by a factor of about 10^{12}. Changes in the spin of the mesotron or the interaction form may reduce this disagreement to 10^{10}.

If the experimental results are correct, they would necessitate a very drastic change in the forms of mesotron interactions. The result is significant also for the production of single mesotrons by artificial sources. Indeed the creation of a mesotron by X-rays or fast protons is the reverse of processes (1) and (2). If the interaction according to these two processes is much weaker than expected, one would conclude the same for the reverse processes. Thus one might be in doubt as to whether one can produce abundant numbers of artificial mesotrons with bombardment-energies only a little above the threshold for single-mesotron production. Predictions concerning the creation of mesotron pairs by electromagnetic radiation are, of course, not affected by these arguments.

† Conversi *et al.*, *loc. cit.*

Index

Selected Readings in Physics

There is a tendency nowadays for undergraduates to learn their physics completely from textbooks, not becoming acquainted with the original literature and therefore not realising how the subject grew and developed. The purpose of the series in which this volume is published is to present a set of reasonably priced books which give, for a particular subject, reprints of those papers which record the development of important new ideas.

KINETIC THEORY

Part A: The Nature of Gases and of Heat

Stephen G. Brush

A.B.(Harvard); D.Phil.(Oxon.)

Lawrence Radiation Laboratory, Livermore, California

Intended as an aid to the teaching of physics from the historical viewpoint, this book includes reprints of selected works on the theory of gases and the nature of heat by Boyle, Newton, Daniel Bernoulli, George Gregory, Robert Mayer, Joule, Helmholtz, Clausius, and Maxwell.

Although for the most part these works expound modern views on the conservation of energy and the kinetic theory of gases, an attempt is made to illustrate the older views that have been replaced by these modern theories. Thus the Boyle–Newton repulsive theory of gases is presented, as well as the mean-free-path explanation of gas viscosity, the Maxwell velocity distribution law, and the virial theorem. The papers selected have been written in a fairly clear and simple style, and should therefore be readily appreciated by the modern reader; they include qualitative descriptions of phenomena and statements of philosophical positions, as well as mathematical derivations. The Introduction provides the proper historical context for the understanding of the reprints.

Selected Readings in Physics

MEN OF PHYSICS: L. D. LANDAU 1

D. ter Haar

L. D. Landau can justly be described as one the world's most eminent physicists. He was awarded the Nobel Prize in Physics and although this was for his work on the theory of condensed media, and especially for his work on the theory of liquid helium, the Prize might equally deservedly have been awarded for other work. Plasma physics, high-energy physics, quantum mechanics, and the theory of magnetism are all topics to which Professor Landau has contributed greatly. Although comprehensively covered in the Collected Papers of L. D. Landau recently published*, it is felt desirable that a selection of some twenty of his most important papers should be made available in the form of two inexpensive publications.

This first volume contains eight papers: two on the theory of helium II, two on the theory of Fermi liquids, two on superconductivity, one on electron diamagnetism and one on ferromagnetism. The second volume contains twelve papers: one on the theory of phase transitions, one on stellar energy, one on the statistical model of nuclei, one on the multiple production of particles in cosmic rays, one on the uncertainty principle in relativistic quantum mechanics, two on the quantum theory of collisions, two on plasma physics, and three on field theory.